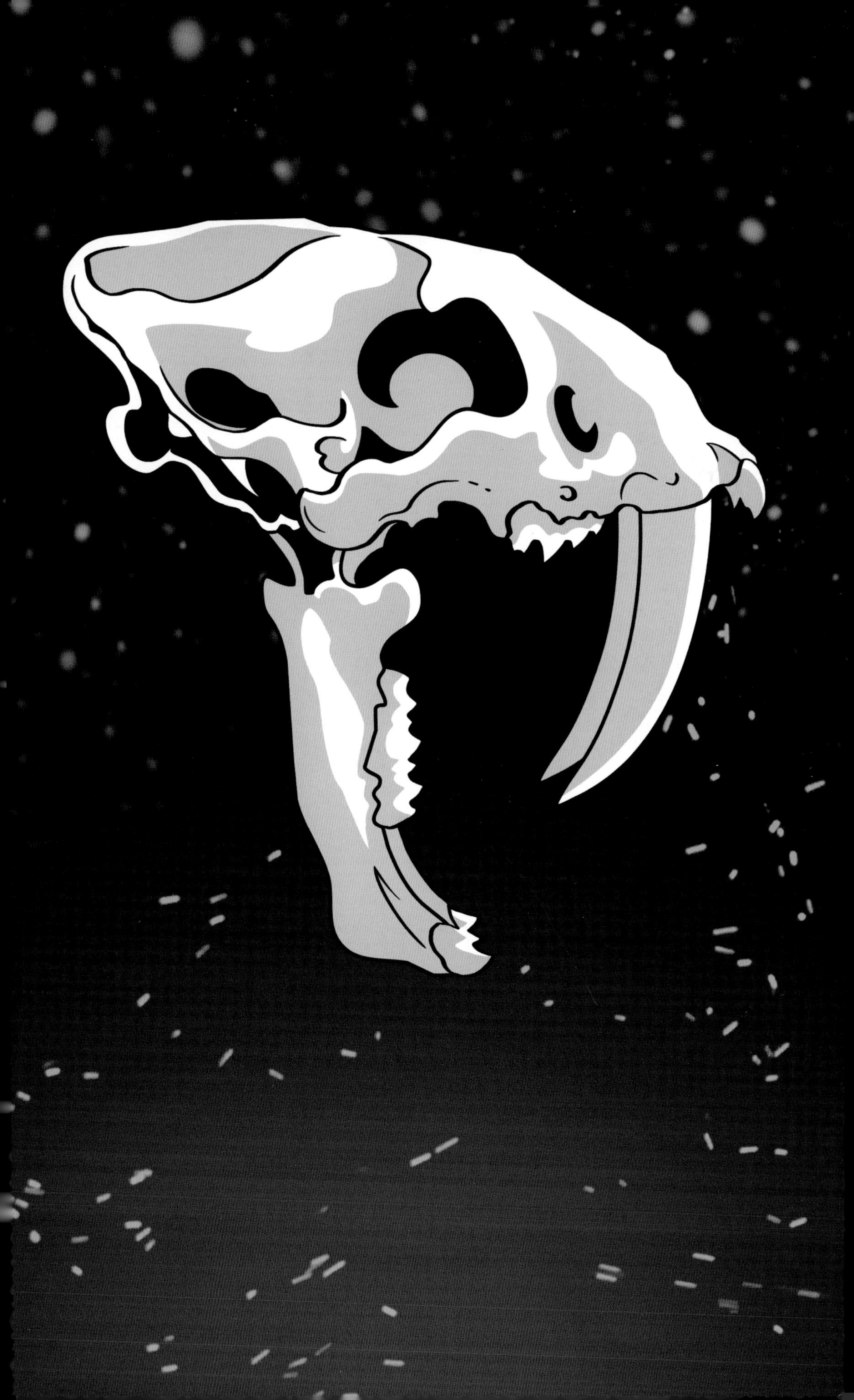

REXTOOTH
STUDIOS

END OF THE
ICE AGE

WRITTEN & ILLUSTRATED BY **TED RECHLIN**

EDITOR **ANNE RECHLIN**

COPYRIGHT © 2018 BY TED RECHLIN

PUBLISHED BY REXTOOTH STUDIOS, BOZEMAN, MONTANA

PRODUCED BY SWEETGRASS BOOKS, HELENA, MONTANA

ISBN: 978-1-59152-220-1

COVER DESIGN BY TED RECHLIN

PRODUCED IN THE UNITED STATES OF AMERICA

PRINTED IN CHINA

Foreword

A world of fantastic beasts existed not so long ago.
From around 2.6 million years ago to 10,000 years ago,
extraordinary creatures roamed our planet.

This is our favorite time in the whole of the Earth's
4.5 billion year history: the Pleistocene
(commonly called the Ice Age).

We are always amazed at the astonishing variety of creatures
that walked the earth at this time, and who wouldn't be?
Gigantic bears as big as a camper van.
Sloths as long as a bus. Enormous cats with canines
as big as steak knives.
Mammoths covered in long, shaggy, hair.
These fabulous beasts might exist
only in myth and imagination – but they were real,
and often overlapped with the dawn of humanity.

Our species, *Homo sapiens*, evolved some
300,000 years ago, and as such our ancestors
witnessed these awesome animals, weaving complex
relationships with them. We hunted them,
drew them on the walls in caves,
and may have even worshipped them.

We have been delighted and privileged to host Ted's work
on our *Twilight Beasts* blog in the past.
Ted's remarkable artwork brings these incredible
creatures to life. Not only do we get to see them
recreated based on the latest scientific knowledge,
we also get a real understanding of the hardships
of life during the very end of the Ice Age as
the climate changed, and with it, the ecosystems
which had supported these amazing animals
for millennia.

Across two continents, Ted brings you into their world,
which was also our world. This book is not just about
some of the unique beasts of the Ice Age.
It is about how we can learn from what we have done
in the past to protect the future.

TWILIGHT BEASTS

www.twilightbeasts.wordpress.com
Jan Freedman, Curator of Natural History, Plymouth Museums Galleries Archives
Rena Maguire, PhD, Iron Age Equestrianism
Ross Barnett, PhD, Ancient DNA

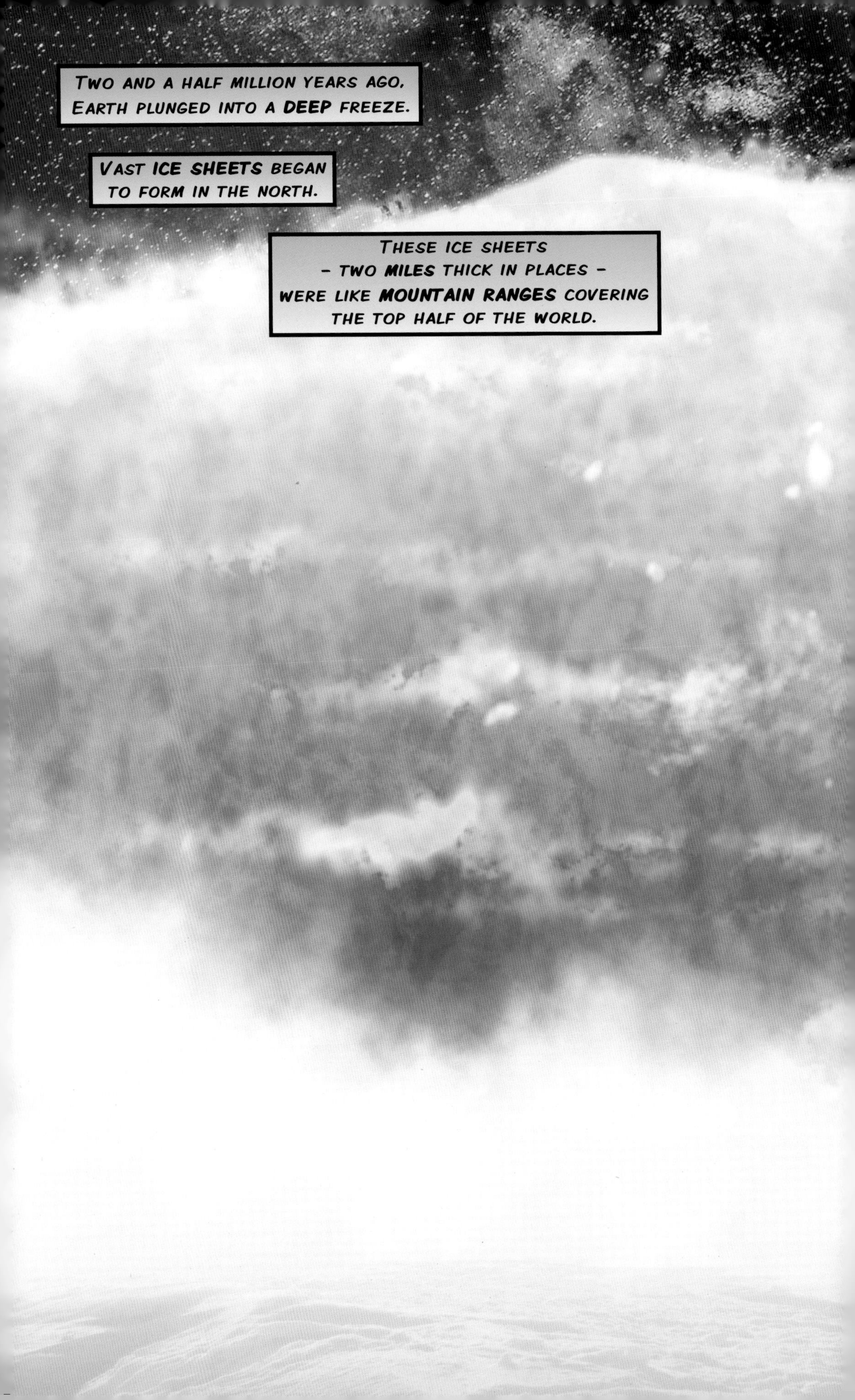
TWO AND A HALF MILLION YEARS AGO, EARTH PLUNGED INTO A DEEP FREEZE.
VAST ICE SHEETS BEGAN TO FORM IN THE NORTH.
THESE ICE SHEETS - TWO MILES THICK IN PLACES - WERE LIKE MOUNTAIN RANGES COVERING THE TOP HALF OF THE WORLD.

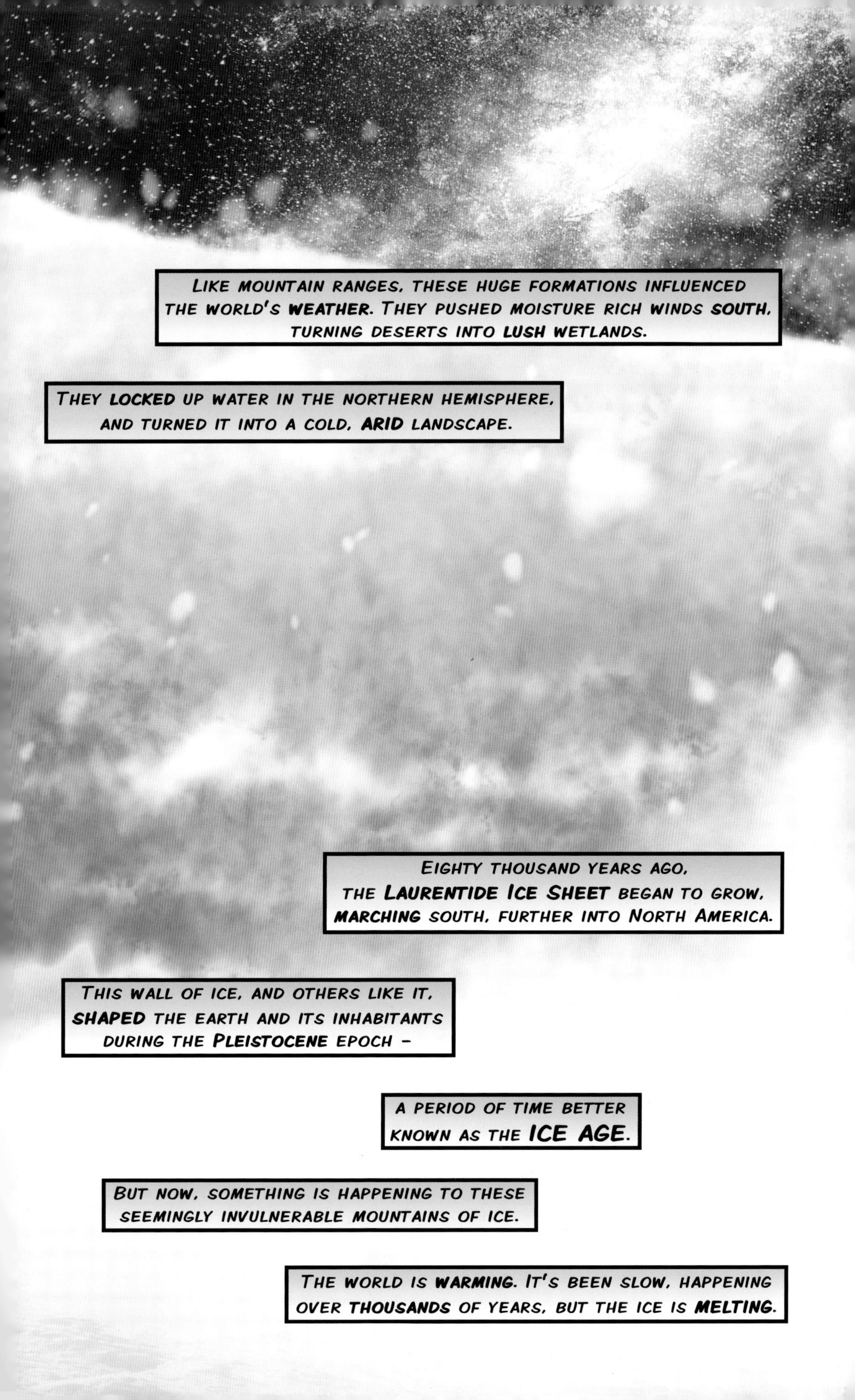
LIKE MOUNTAIN RANGES, THESE HUGE FORMATIONS INFLUENCED THE WORLD'S WEATHER. THEY PUSHED MOISTURE RICH WINDS SOUTH, TURNING DESERTS INTO LUSH WETLANDS.
THEY LOCKED UP WATER IN THE NORTHERN HEMISPHERE, AND TURNED IT INTO A COLD, ARID LANDSCAPE.
EIGHTY THOUSAND YEARS AGO, THE LAURENTIDE ICE SHEET BEGAN TO GROW, MARCHING SOUTH, FURTHER INTO NORTH AMERICA.
THIS WALL OF ICE, AND OTHERS LIKE IT, SHAPED THE EARTH AND ITS INHABITANTS DURING THE PLEISTOCENE EPOCH –
A PERIOD OF TIME BETTER KNOWN AS THE ICE AGE.
BUT NOW, SOMETHING IS HAPPENING TO THESE SEEMINGLY INVULNERABLE MOUNTAINS OF ICE.
THE WORLD IS WARMING. IT'S BEEN SLOW, HAPPENING OVER THOUSANDS OF YEARS, BUT THE ICE IS MELTING.

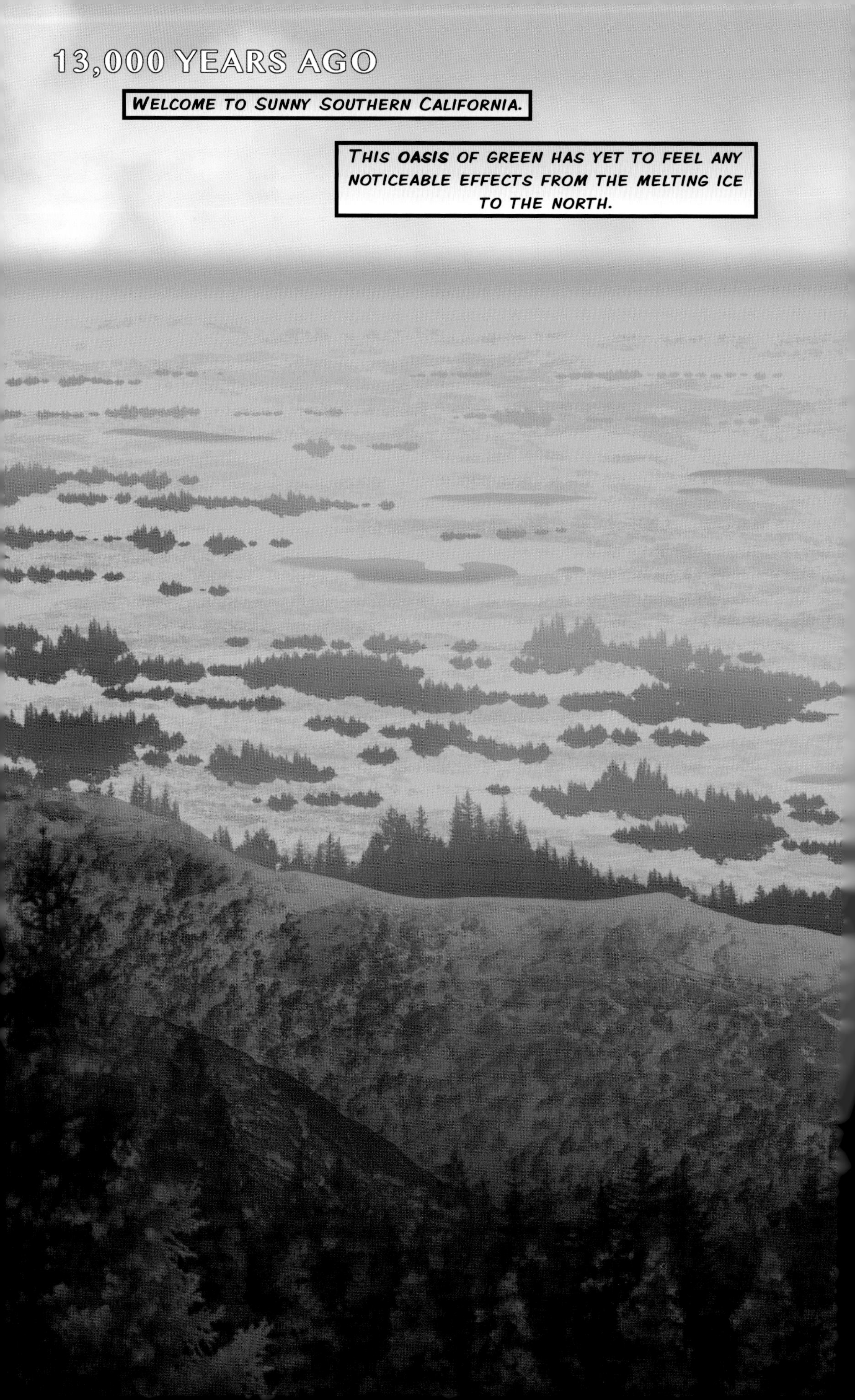
13,000 YEARS AGO
WELCOME TO SUNNY SOUTHERN CALIFORNIA.
THIS OASIS OF GREEN HAS YET TO FEEL ANY NOTICEABLE EFFECTS FROM THE MELTING ICE TO THE NORTH.

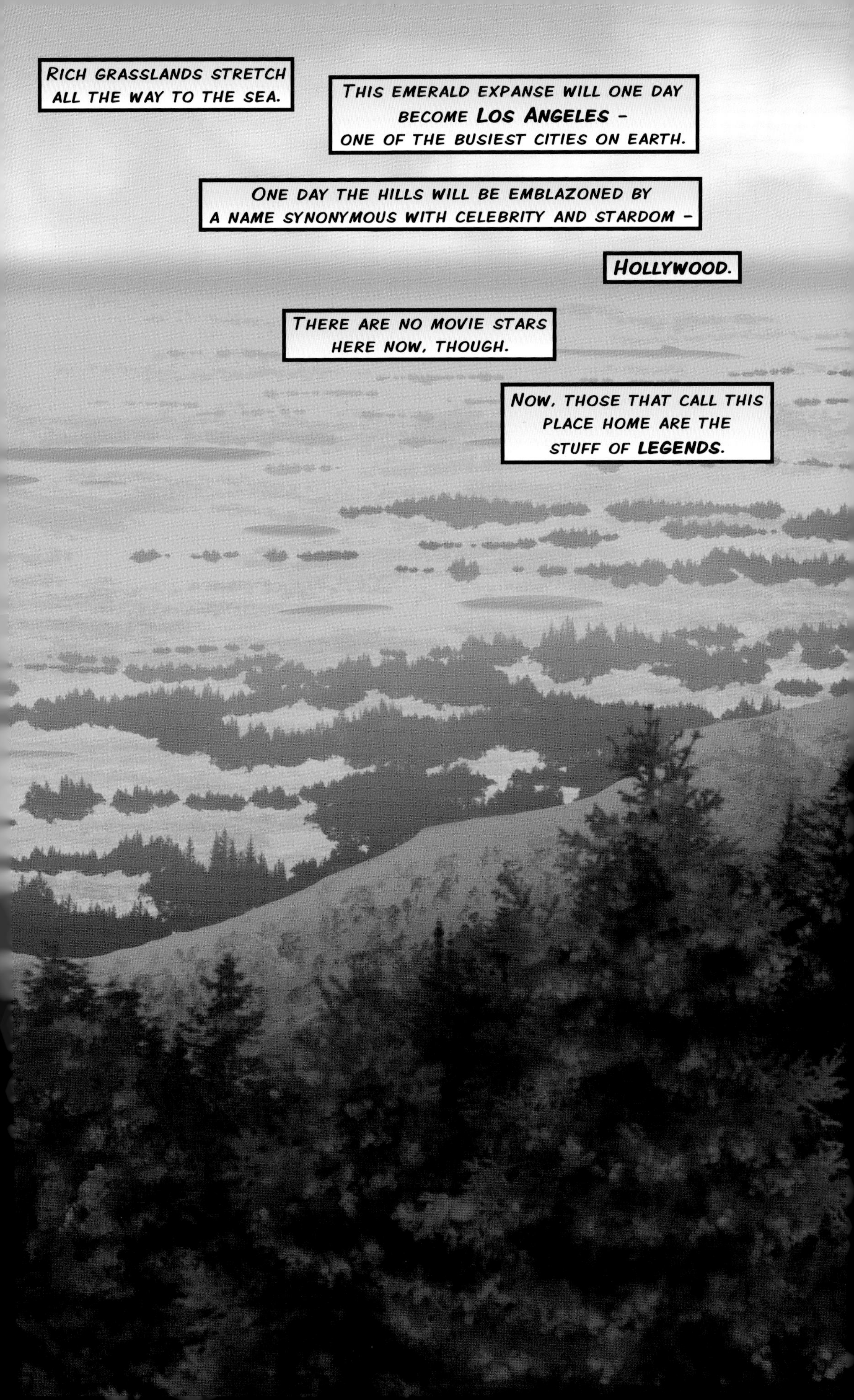
Rich grasslands stretch all the way to the sea.
This emerald expanse will one day become Los Angeles – one of the busiest cities on earth.
One day the hills will be emblazoned by a name synonymous with celebrity and stardom –
Hollywood.
There are no movie stars here now, though.
Now, those that call this place home are the stuff of legends.

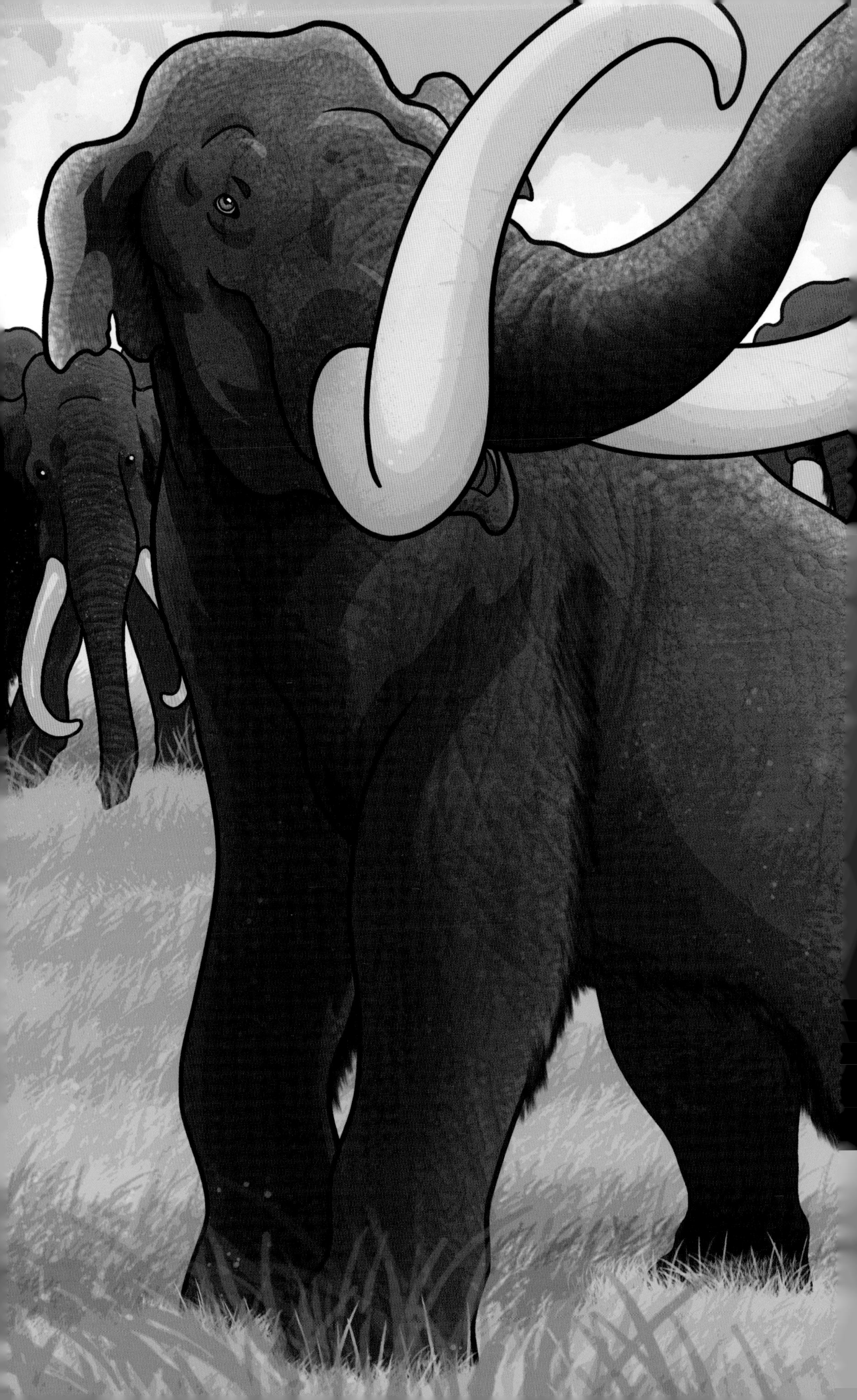

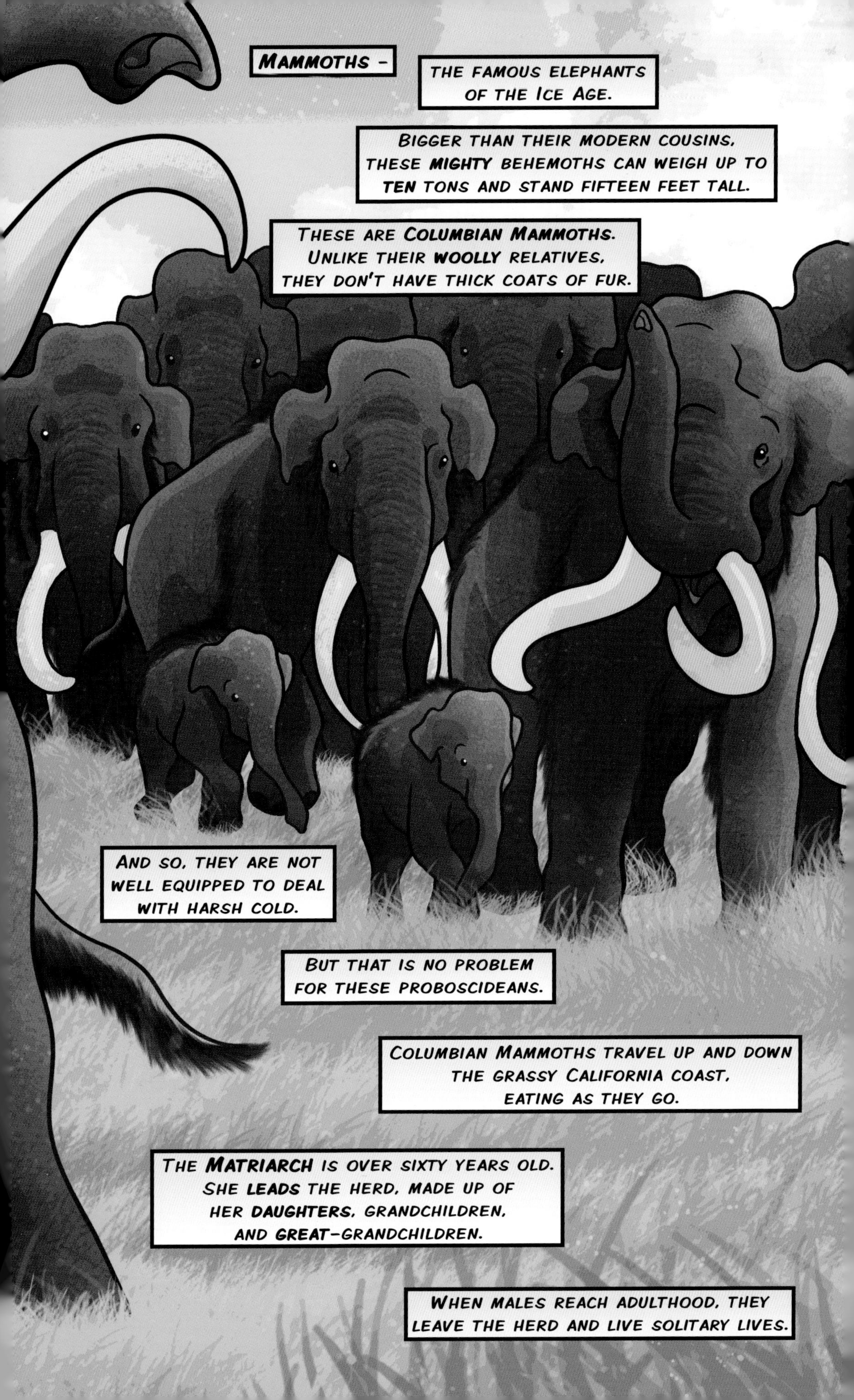
MAMMOTHS -
THE FAMOUS ELEPHANTS OF THE ICE AGE.
BIGGER THAN THEIR MODERN COUSINS, THESE **MIGHTY** BEHEMOTHS CAN WEIGH UP TO **TEN** TONS AND STAND FIFTEEN FEET TALL.
THESE ARE **COLUMBIAN MAMMOTHS**. UNLIKE THEIR **WOOLLY** RELATIVES, THEY DON'T HAVE THICK COATS OF FUR.
AND SO, THEY ARE NOT WELL EQUIPPED TO DEAL WITH HARSH COLD.
BUT THAT IS NO PROBLEM FOR THESE PROBOSCIDEANS.
COLUMBIAN MAMMOTHS TRAVEL UP AND DOWN THE GRASSY CALIFORNIA COAST, EATING AS THEY GO.
THE **MATRIARCH** IS OVER SIXTY YEARS OLD. SHE **LEADS** THE HERD, MADE UP OF HER **DAUGHTERS**, GRANDCHILDREN, AND **GREAT**-GRANDCHILDREN.
WHEN MALES REACH ADULTHOOD, THEY LEAVE THE HERD AND LIVE SOLITARY LIVES.

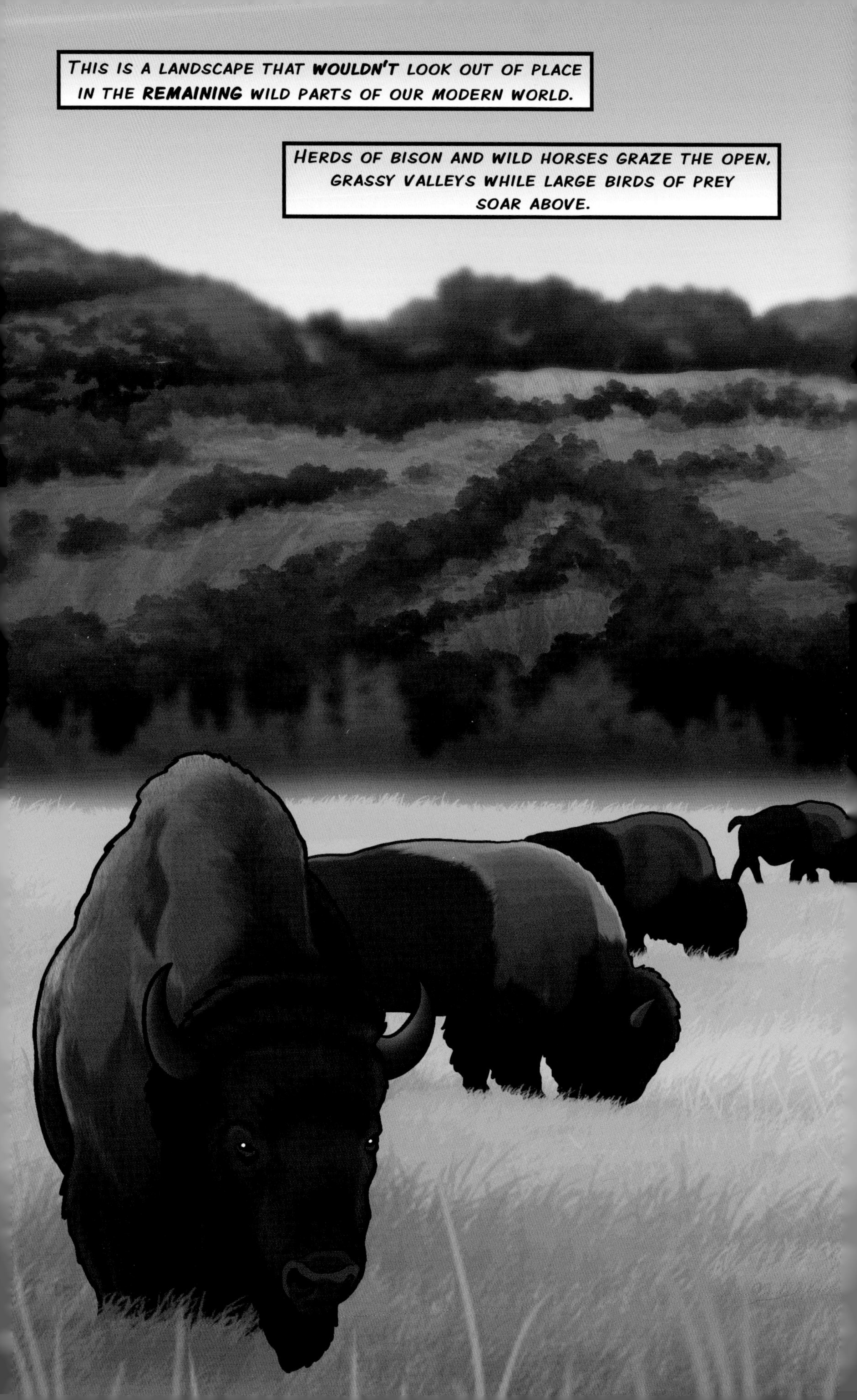
THIS IS A LANDSCAPE THAT **WOULDN'T** LOOK OUT OF PLACE IN THE **REMAINING** WILD PARTS OF OUR MODERN WORLD.
HERDS OF BISON AND WILD HORSES GRAZE THE OPEN, GRASSY VALLEYS WHILE LARGE BIRDS OF PREY SOAR ABOVE.

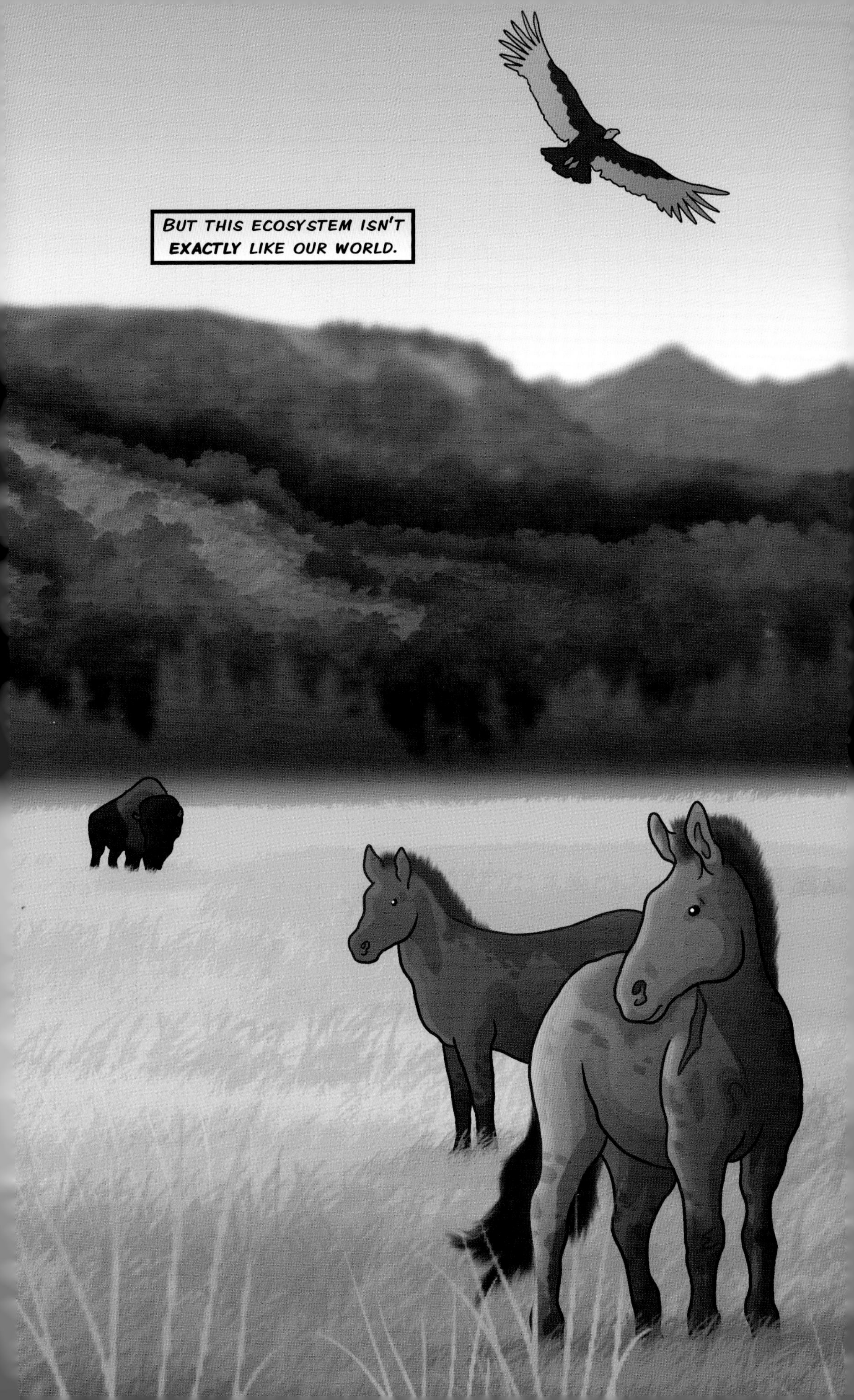
BUT THIS ECOSYSTEM ISN'T
EXACTLY LIKE OUR WORLD.

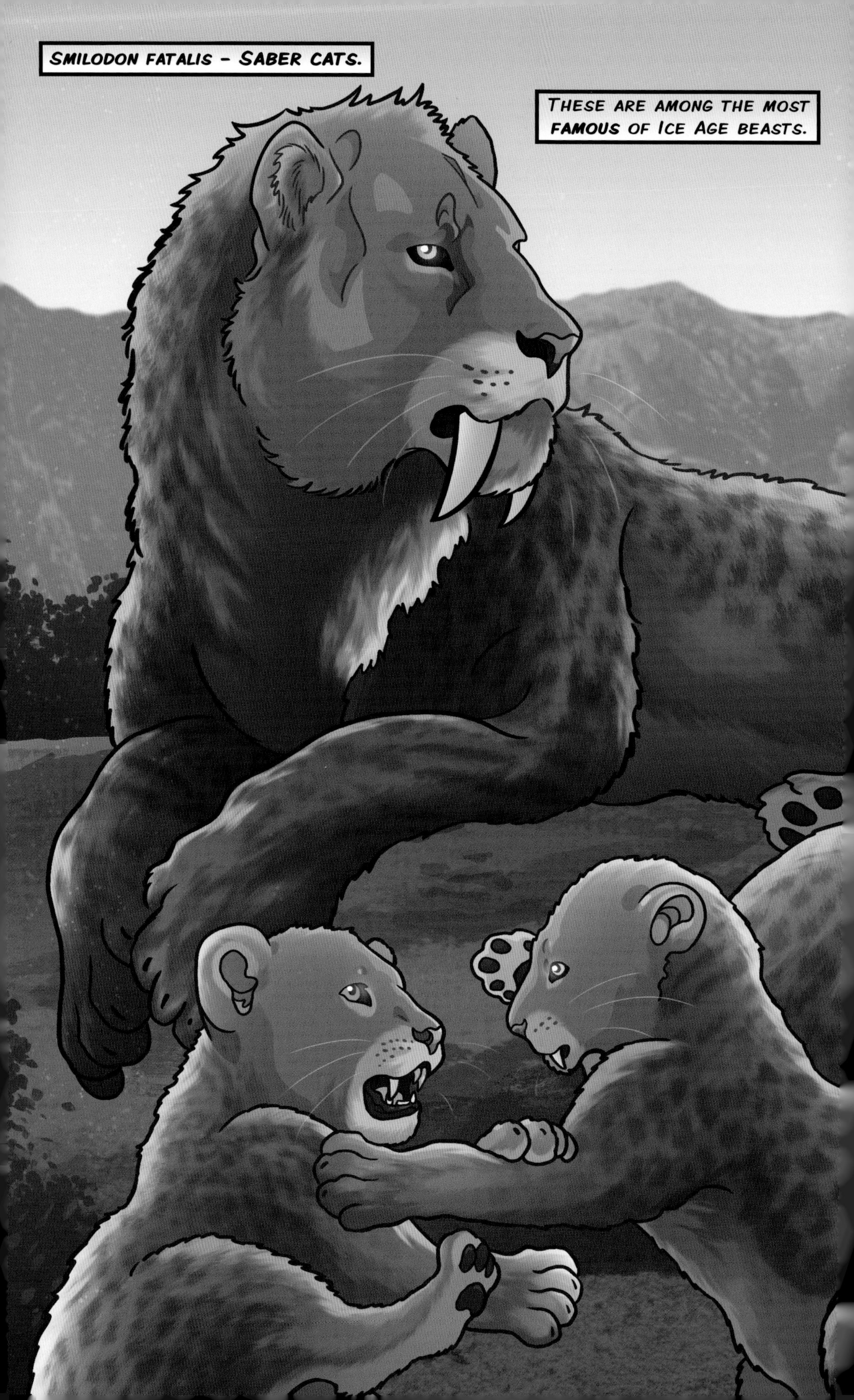
SMILODON FATALIS – SABER CATS.
THESE ARE AMONG THE MOST FAMOUS OF ICE AGE BEASTS.

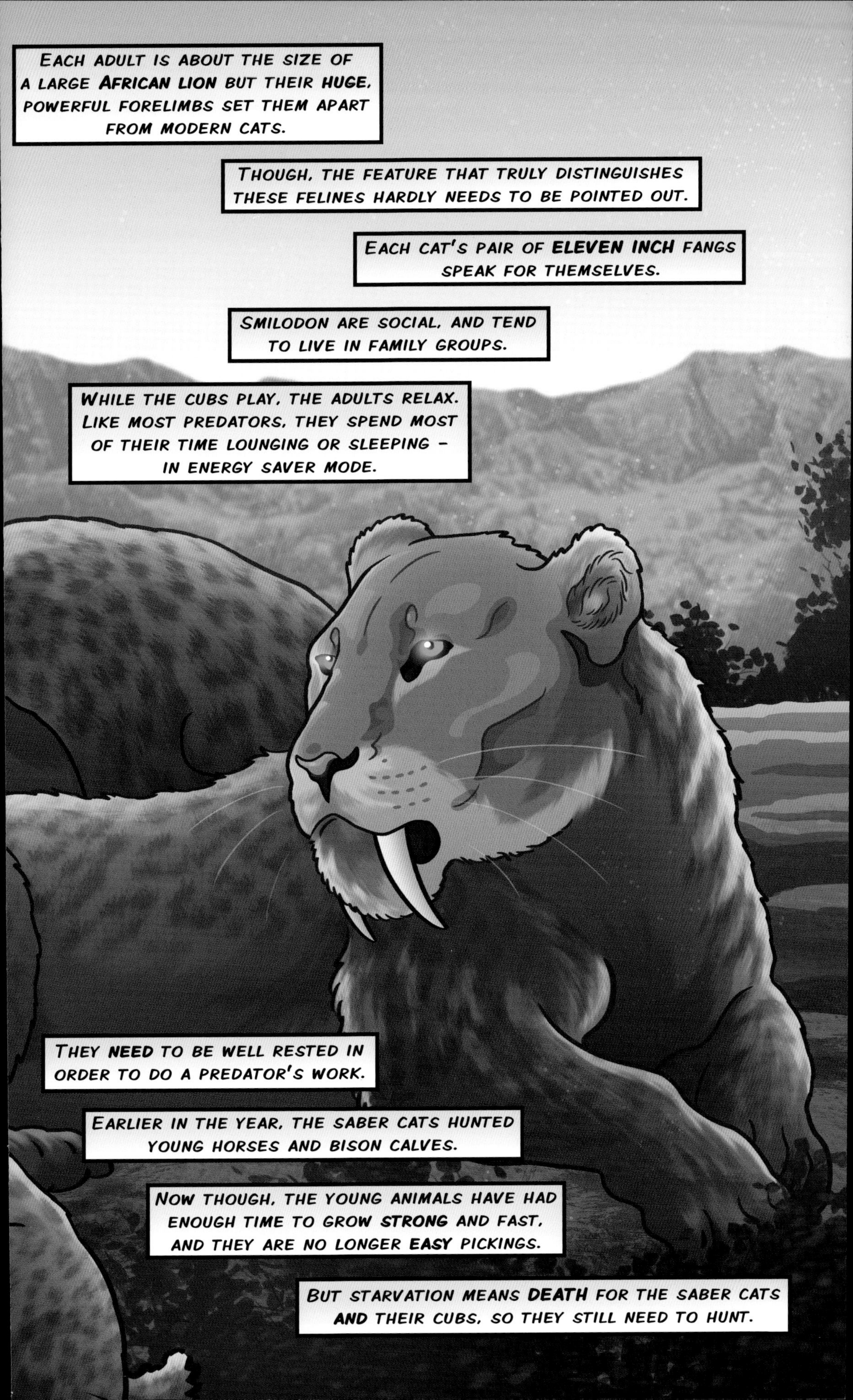
EACH ADULT IS ABOUT THE SIZE OF A LARGE **AFRICAN LION** BUT THEIR **HUGE**, POWERFUL FORELIMBS SET THEM APART FROM MODERN CATS.
THOUGH, THE FEATURE THAT TRULY DISTINGUISHES THESE FELINES HARDLY NEEDS TO BE POINTED OUT.
EACH CAT'S PAIR OF **ELEVEN INCH** FANGS SPEAK FOR THEMSELVES.
SMILODON ARE SOCIAL, AND TEND TO LIVE IN FAMILY GROUPS.
WHILE THE CUBS PLAY, THE ADULTS RELAX. LIKE MOST PREDATORS, THEY SPEND MOST OF THEIR TIME LOUNGING OR SLEEPING – IN ENERGY SAVER MODE.
THEY **NEED** TO BE WELL RESTED IN ORDER TO DO A PREDATOR'S WORK.
EARLIER IN THE YEAR, THE SABER CATS HUNTED YOUNG HORSES AND BISON CALVES.
NOW THOUGH, THE YOUNG ANIMALS HAVE HAD ENOUGH TIME TO GROW **STRONG** AND FAST, AND THEY ARE NO LONGER **EASY** PICKINGS.
BUT STARVATION MEANS **DEATH** FOR THE SABER CATS **AND** THEIR CUBS, SO THEY STILL NEED TO HUNT.

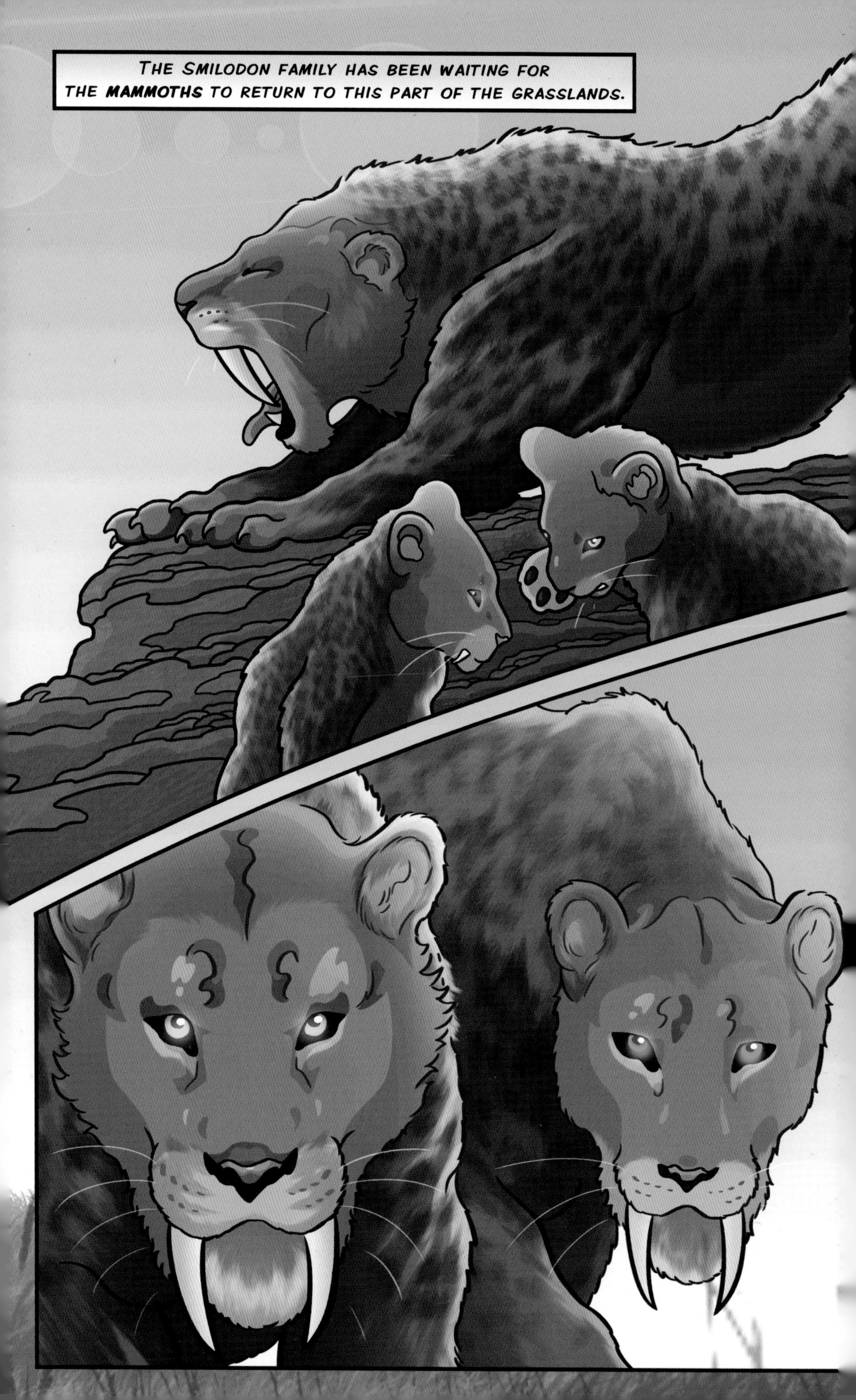
The Smilodon family has been waiting for the **mammoths** to return to this part of the grasslands.

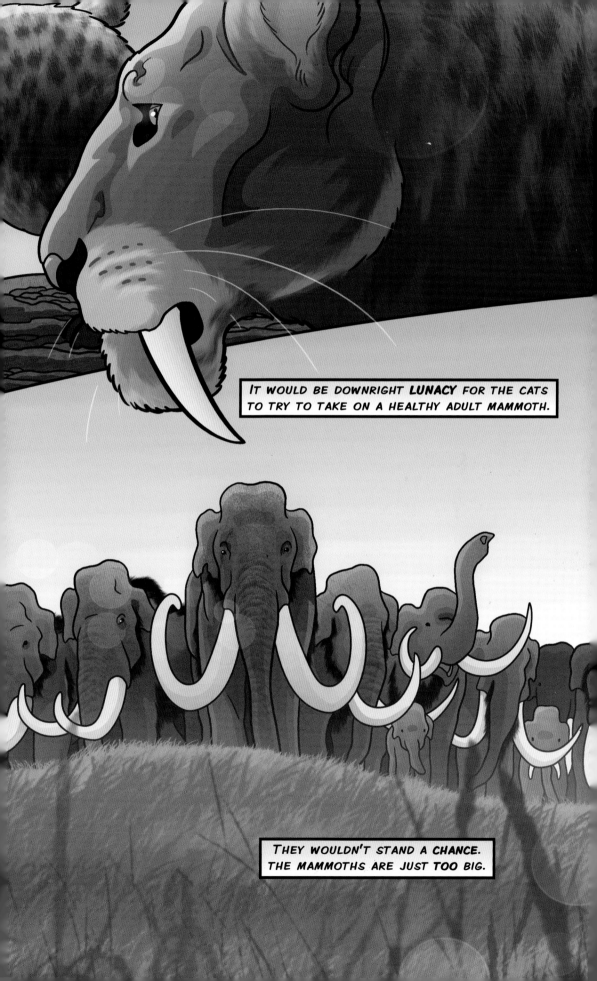

It would be downright **LUNACY** for the cats to try to take on a healthy adult mammoth.

They wouldn't stand a **chance.** The mammoths are just **too big.**

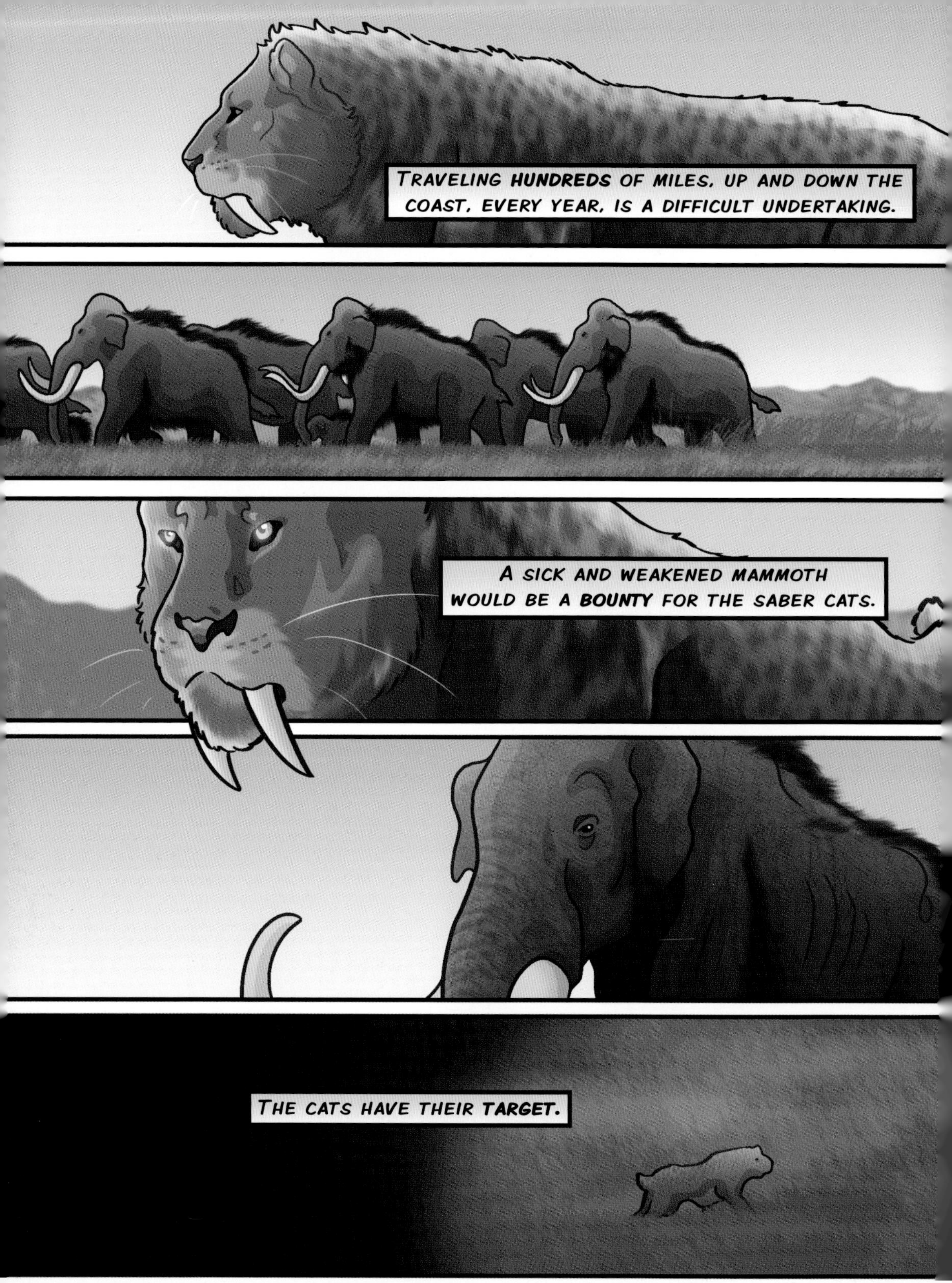
TRAVELING HUNDREDS OF MILES, UP AND DOWN THE COAST, EVERY YEAR, IS A DIFFICULT UNDERTAKING.
A SICK AND WEAKENED MAMMOTH WOULD BE A BOUNTY FOR THE SABER CATS.
THE CATS HAVE THEIR TARGET.

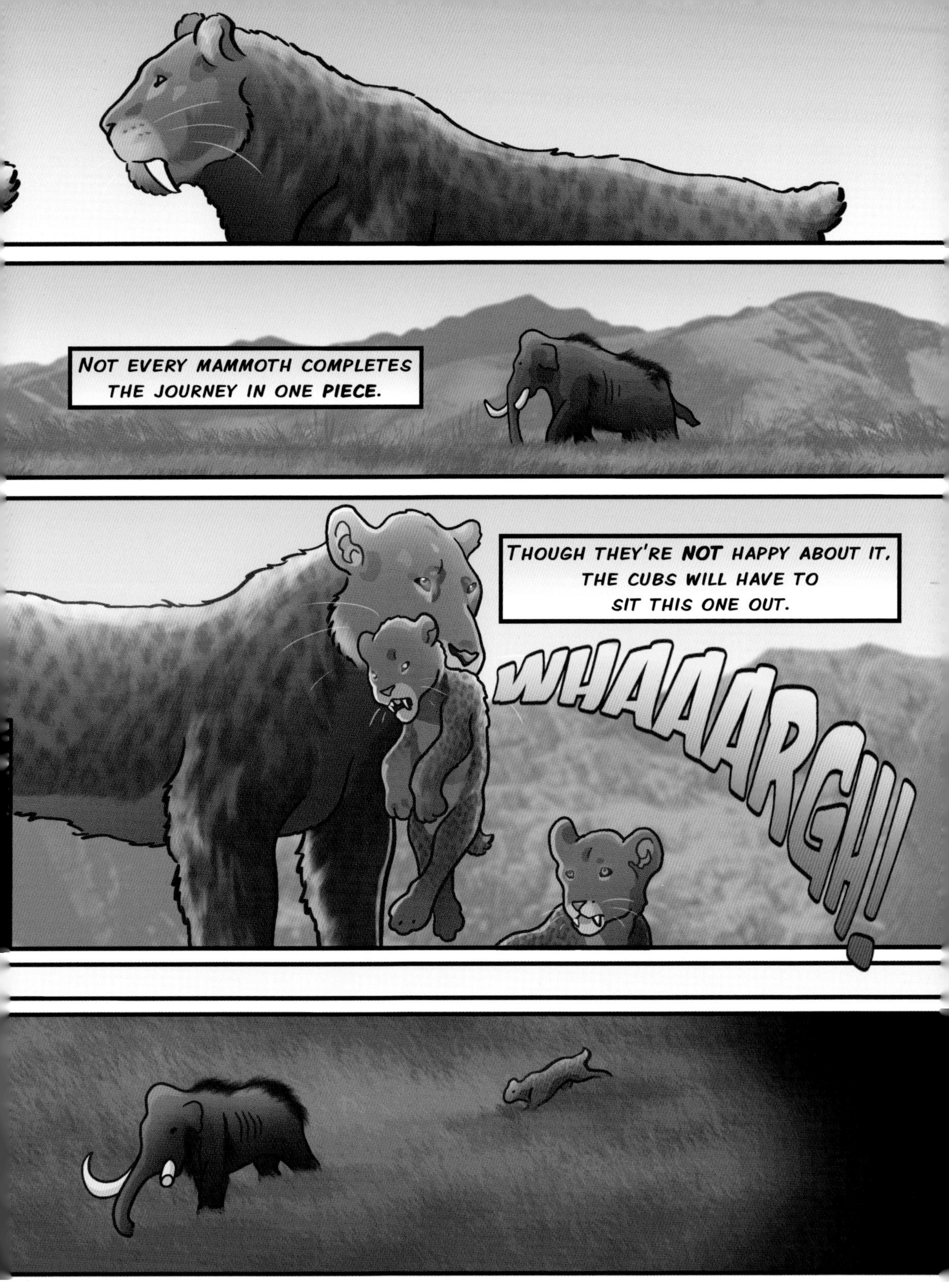
NOT EVERY MAMMOTH COMPLETES THE JOURNEY IN ONE PIECE.
THOUGH THEY'RE NOT HAPPY ABOUT IT, THE CUBS WILL HAVE TO SIT THIS ONE OUT.
WHAAAARGH!!

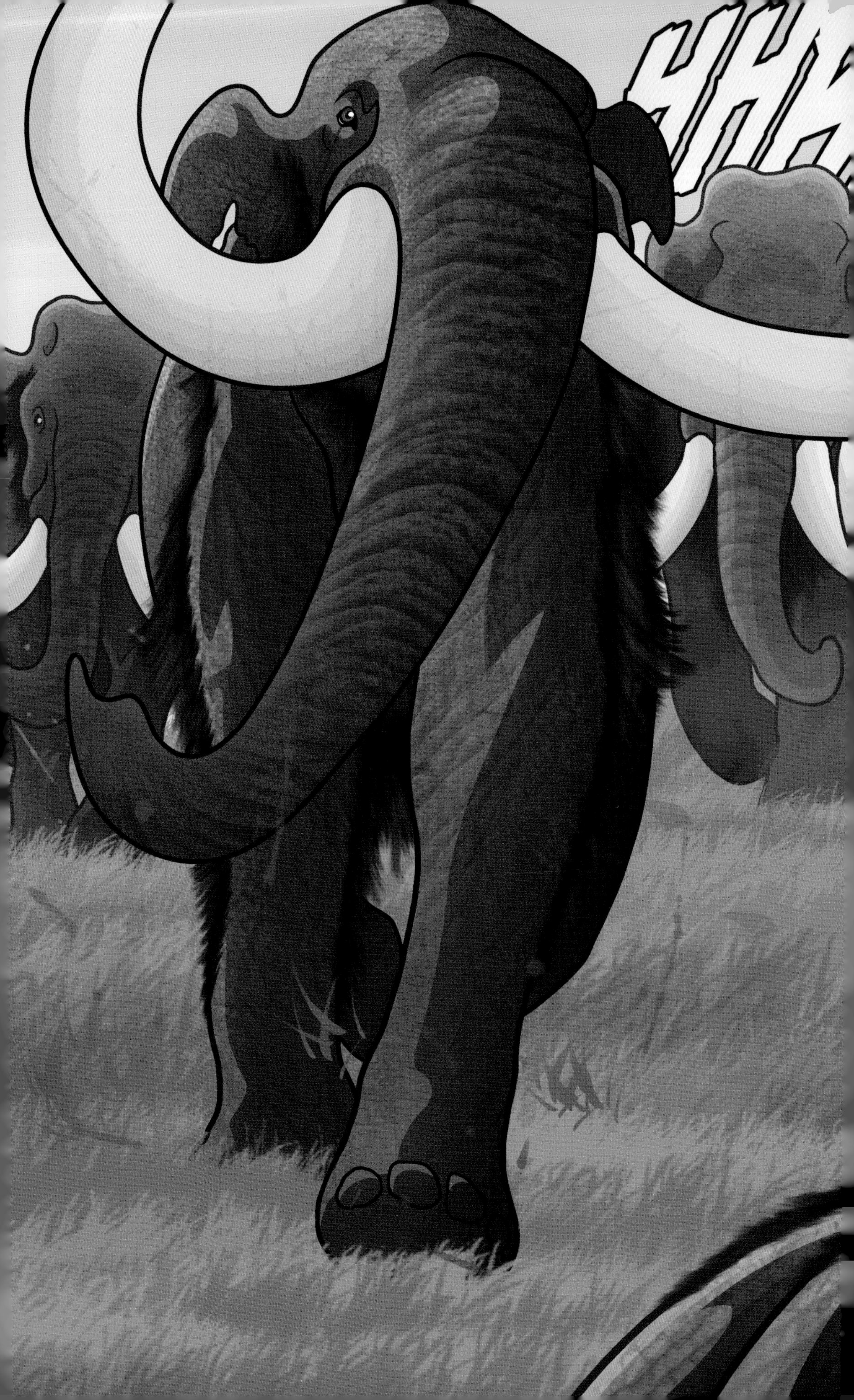

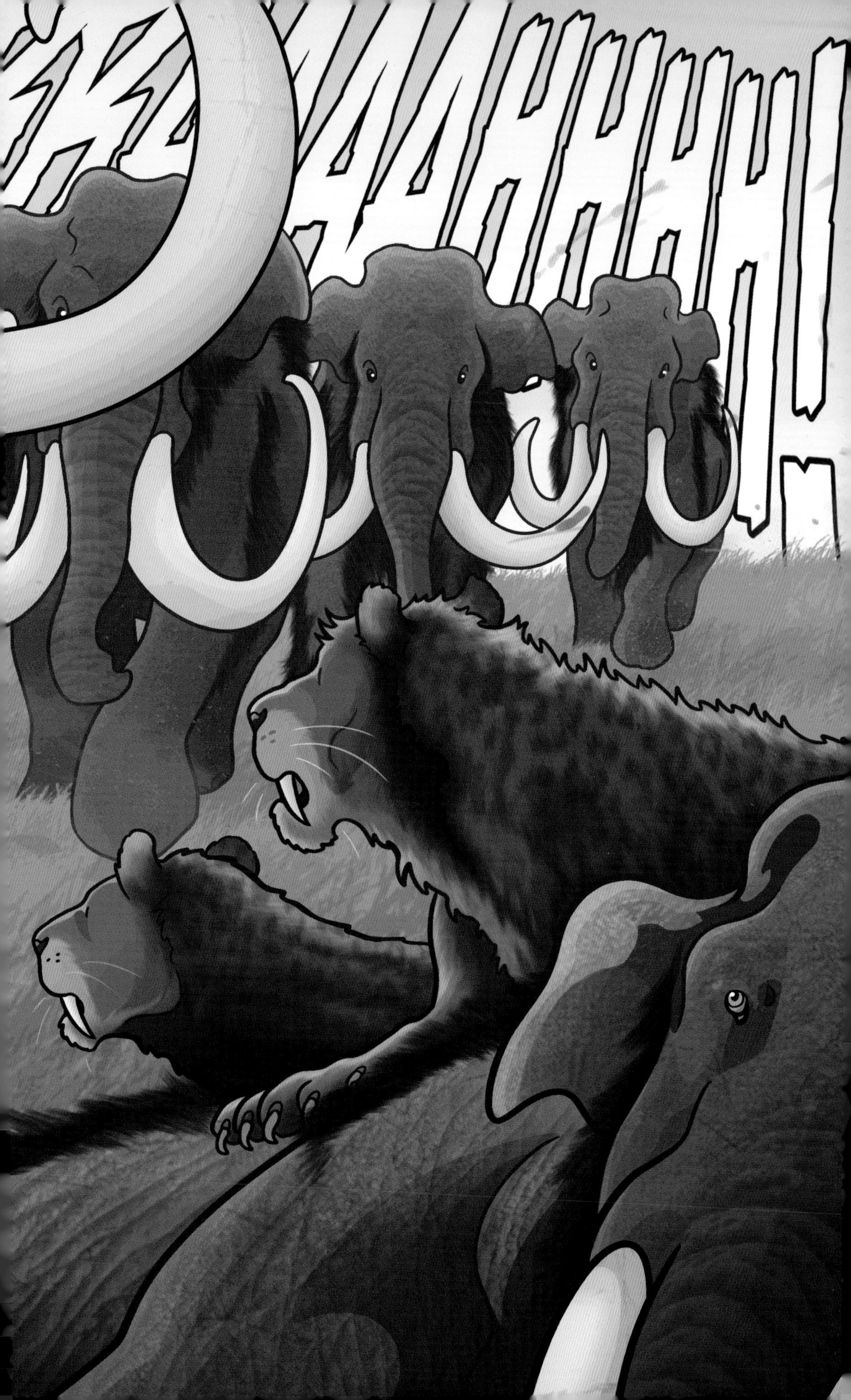

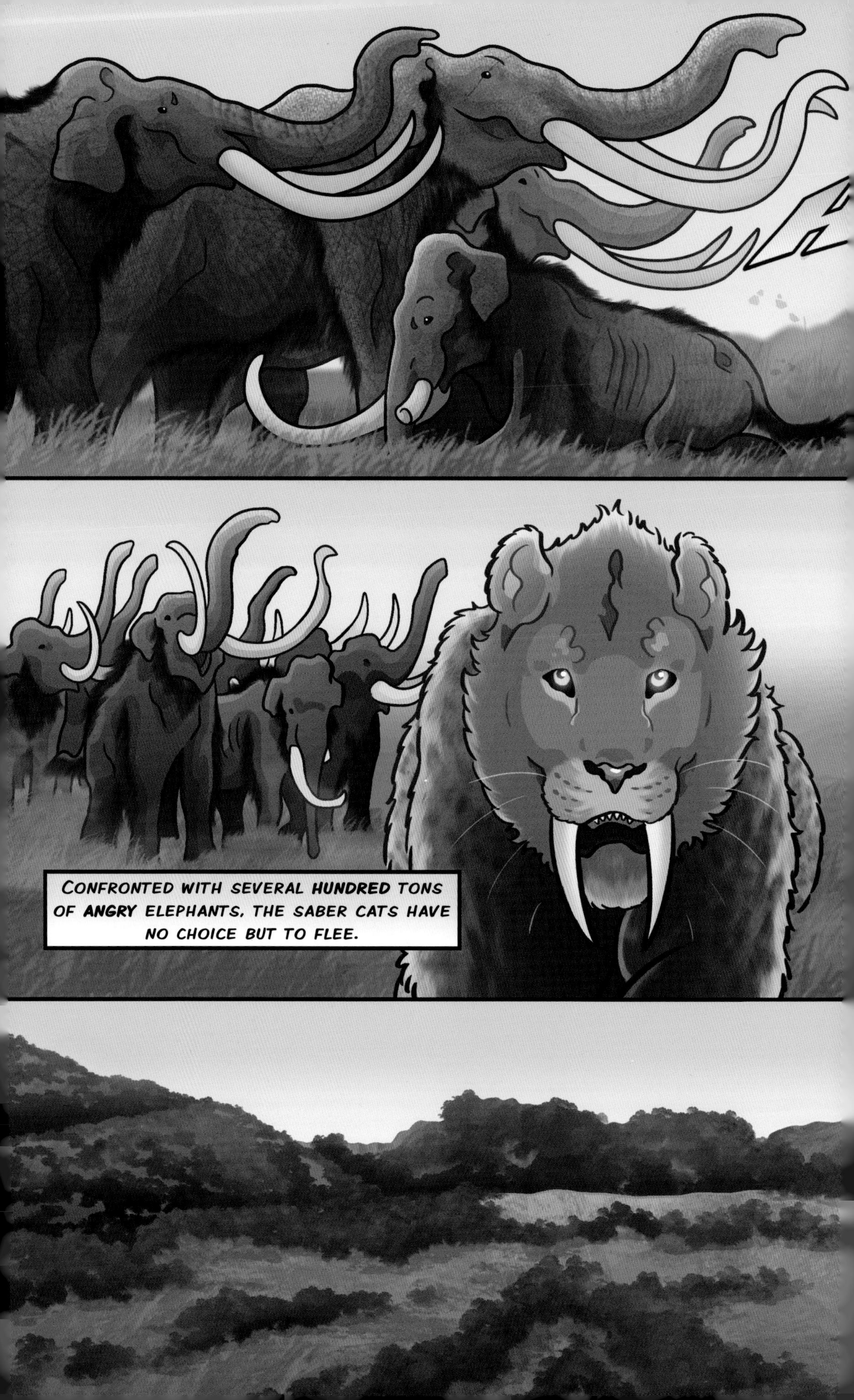
CONFRONTED WITH SEVERAL HUNDRED TONS OF ANGRY ELEPHANTS, THE SABER CATS HAVE NO CHOICE BUT TO FLEE.

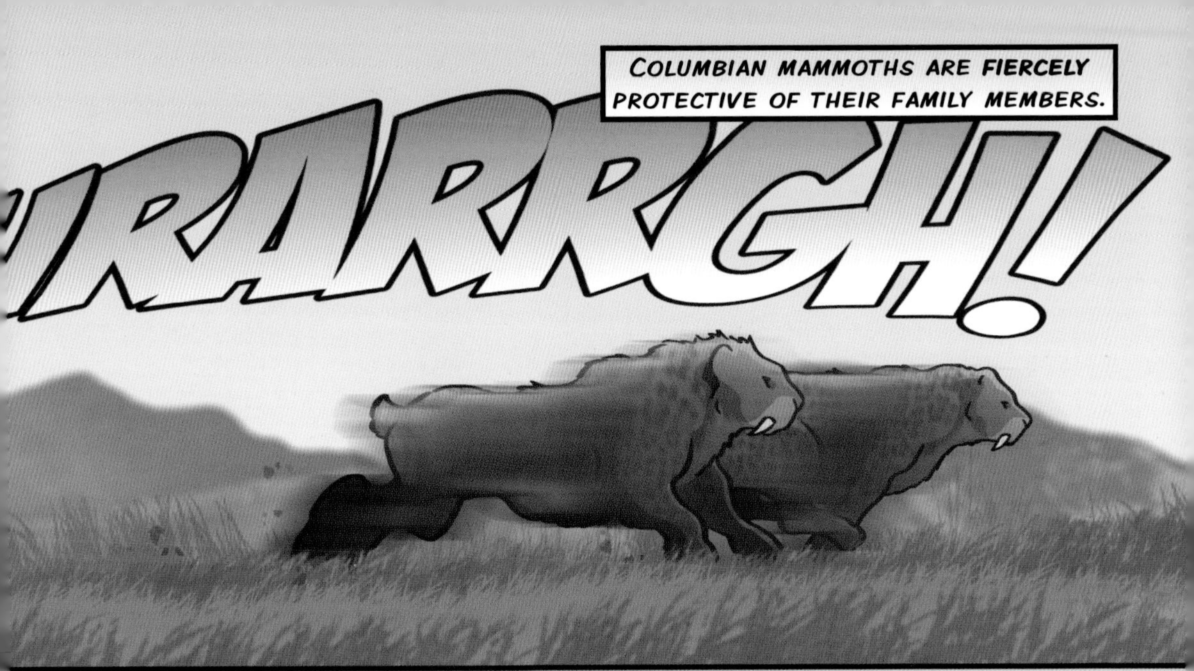
COLUMBIAN MAMMOTHS ARE FIERCELY PROTECTIVE OF THEIR FAMILY MEMBERS.
RARRGH!

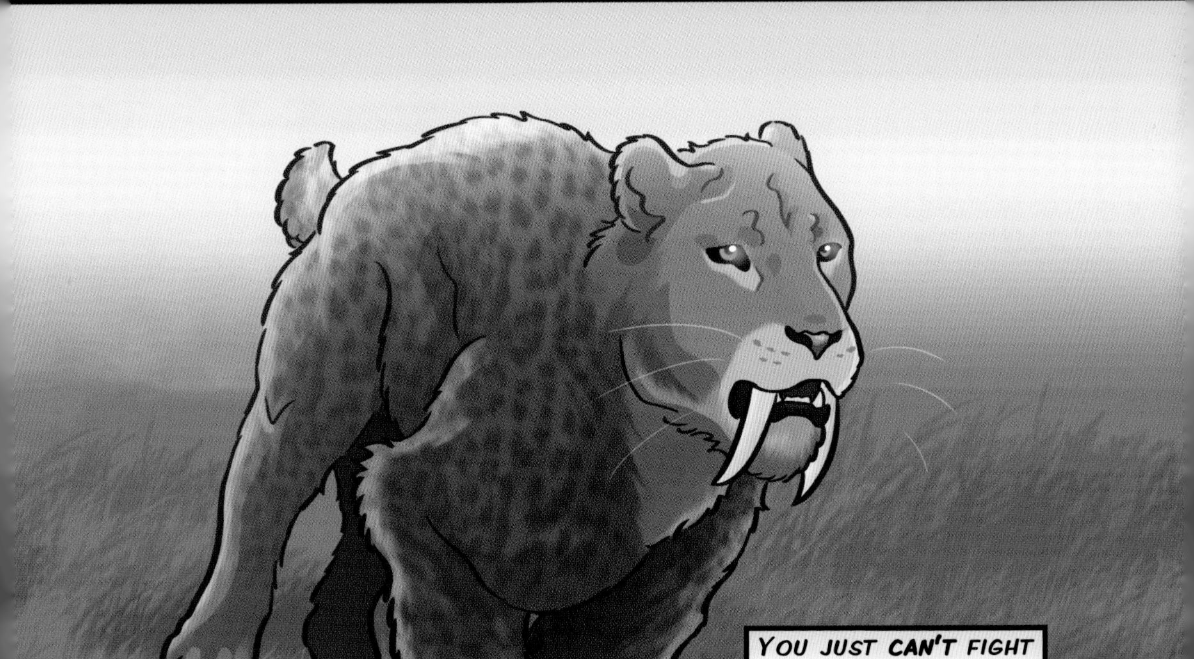
YOU JUST CAN'T FIGHT THE HERD.

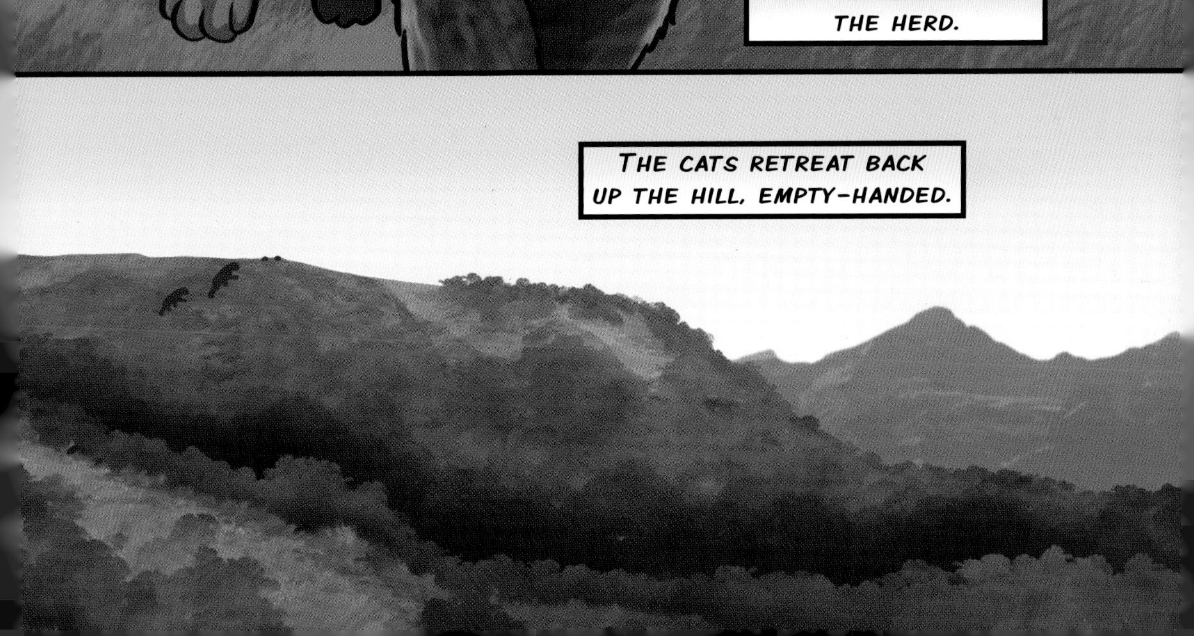
THE CATS RETREAT BACK UP THE HILL, EMPTY-HANDED.

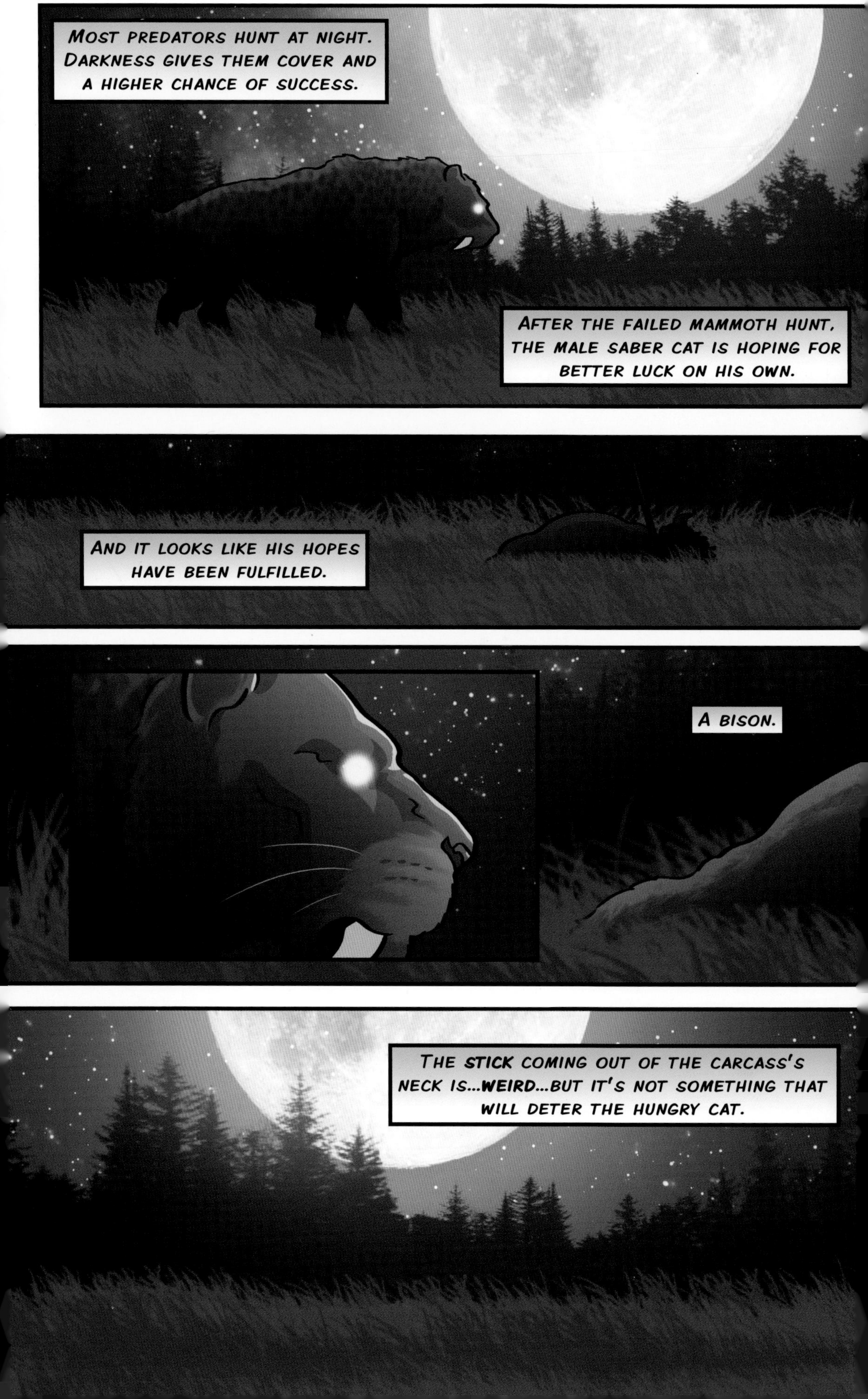
MOST PREDATORS HUNT AT NIGHT. DARKNESS GIVES THEM COVER AND A HIGHER CHANCE OF SUCCESS.
AFTER THE FAILED MAMMOTH HUNT, THE MALE SABER CAT IS HOPING FOR BETTER LUCK ON HIS OWN.
AND IT LOOKS LIKE HIS HOPES HAVE BEEN FULFILLED.
A BISON.
THE STICK COMING OUT OF THE CARCASS'S NECK IS...WEIRD...BUT IT'S NOT SOMETHING THAT WILL DETER THE HUNGRY CAT.

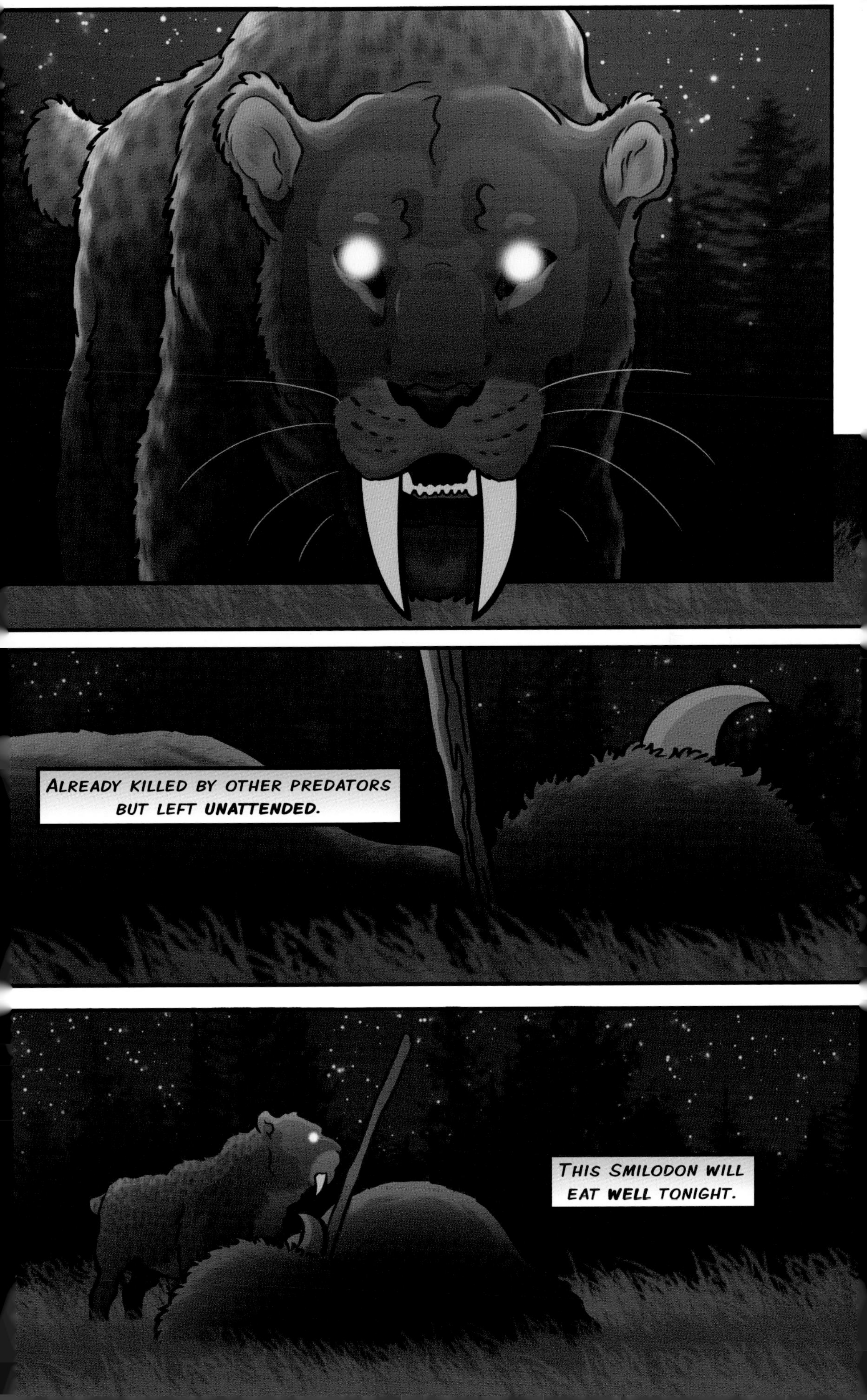
ALREADY KILLED BY OTHER PREDATORS BUT LEFT UNATTENDED.
THIS SMILODON WILL EAT WELL TONIGHT.

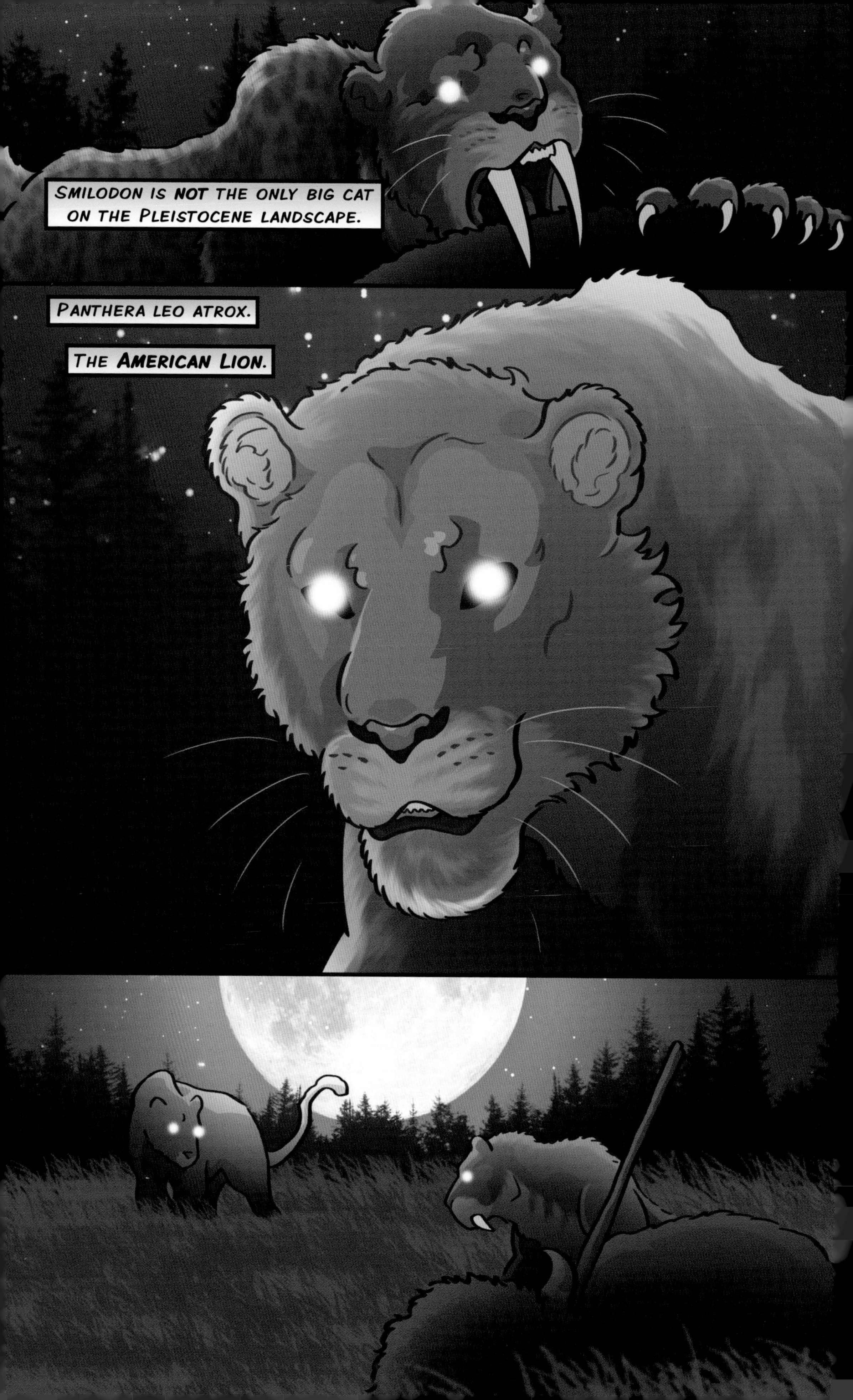

SMILODON IS NOT THE ONLY BIG CAT ON THE PLEISTOCENE LANDSCAPE.
PANTHERA LEO ATROX.
THE AMERICAN LION.

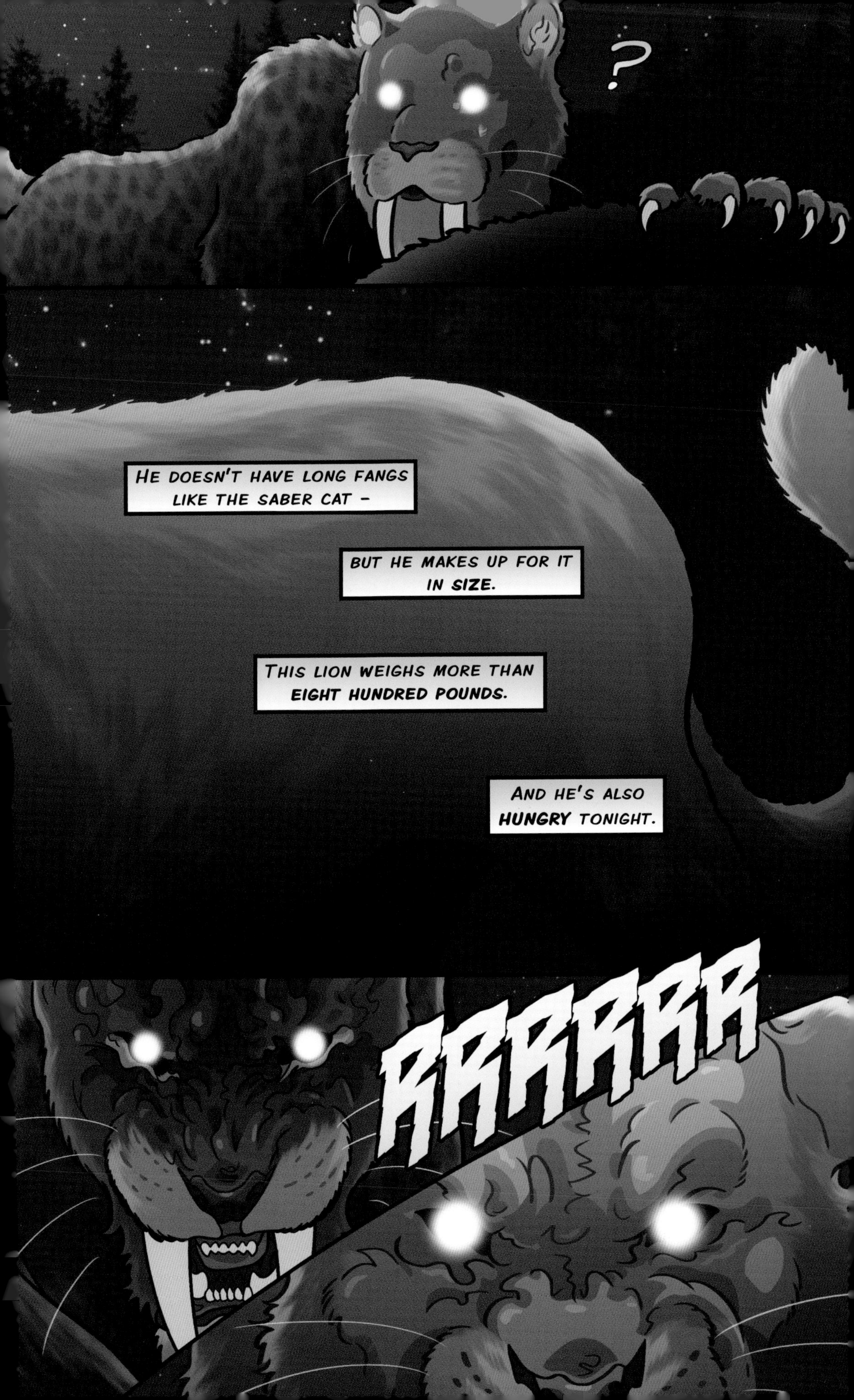
?
HE DOESN'T HAVE LONG FANGS LIKE THE SABER CAT –
BUT HE MAKES UP FOR IT IN SIZE.
THIS LION WEIGHS MORE THAN EIGHT HUNDRED POUNDS.
AND HE'S ALSO HUNGRY TONIGHT.
RRRRRR

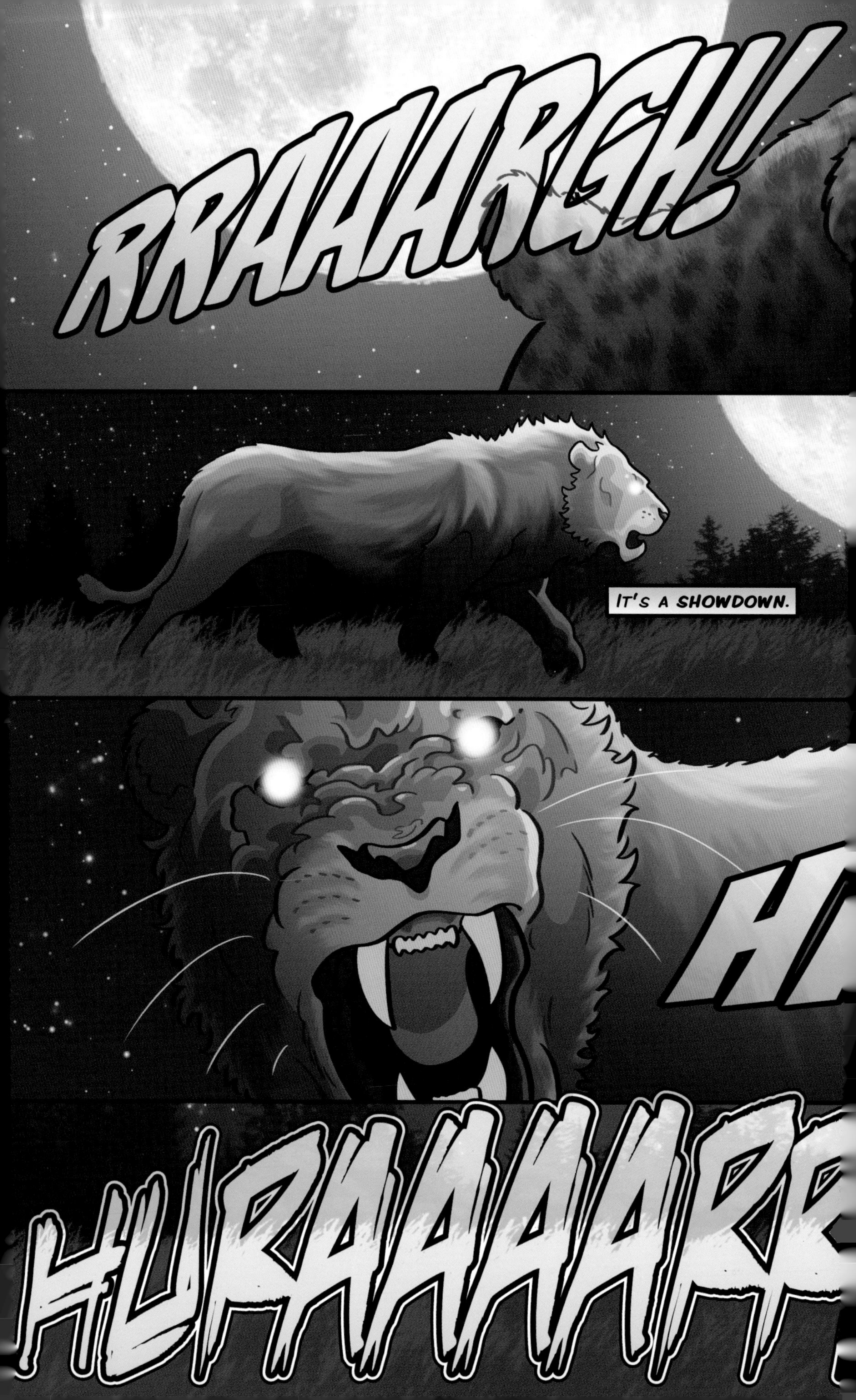
RRAAARGH!
IT'S A SHOWDOWN.

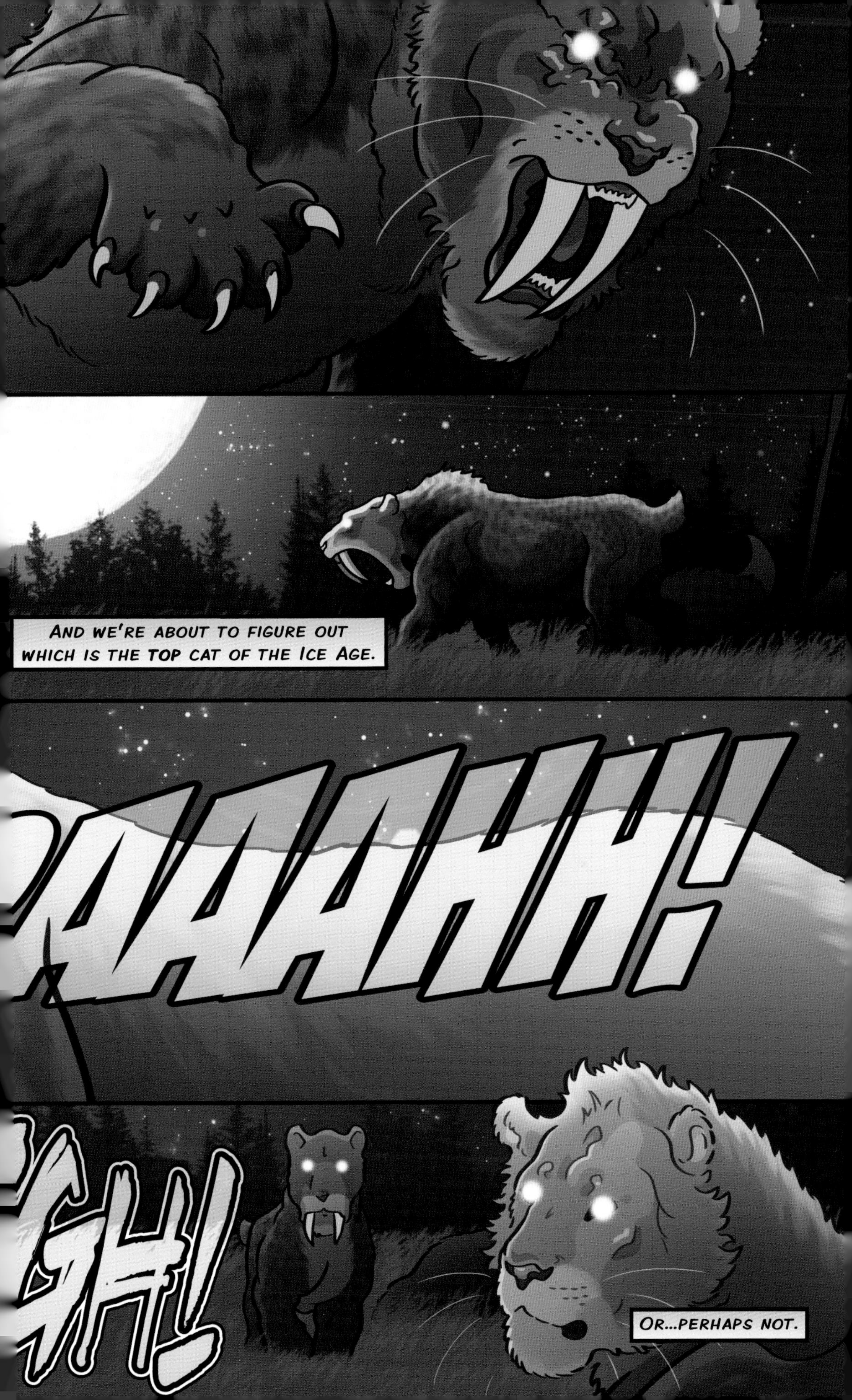
AND WE'RE ABOUT TO FIGURE OUT WHICH IS THE TOP CAT OF THE ICE AGE.
OR...PERHAPS NOT.

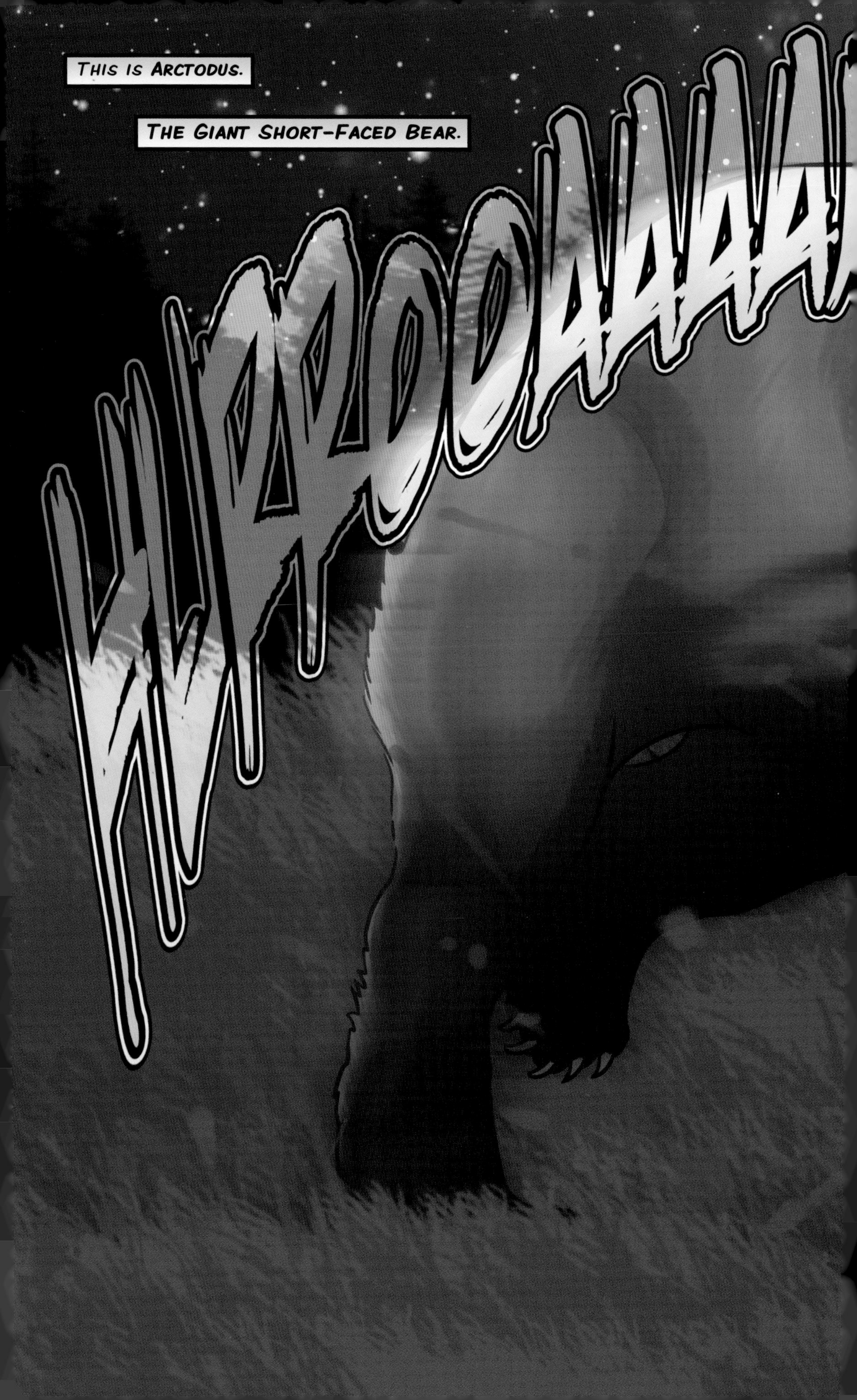
THIS IS ARCTODUS.
THE GIANT SHORT-FACED BEAR.

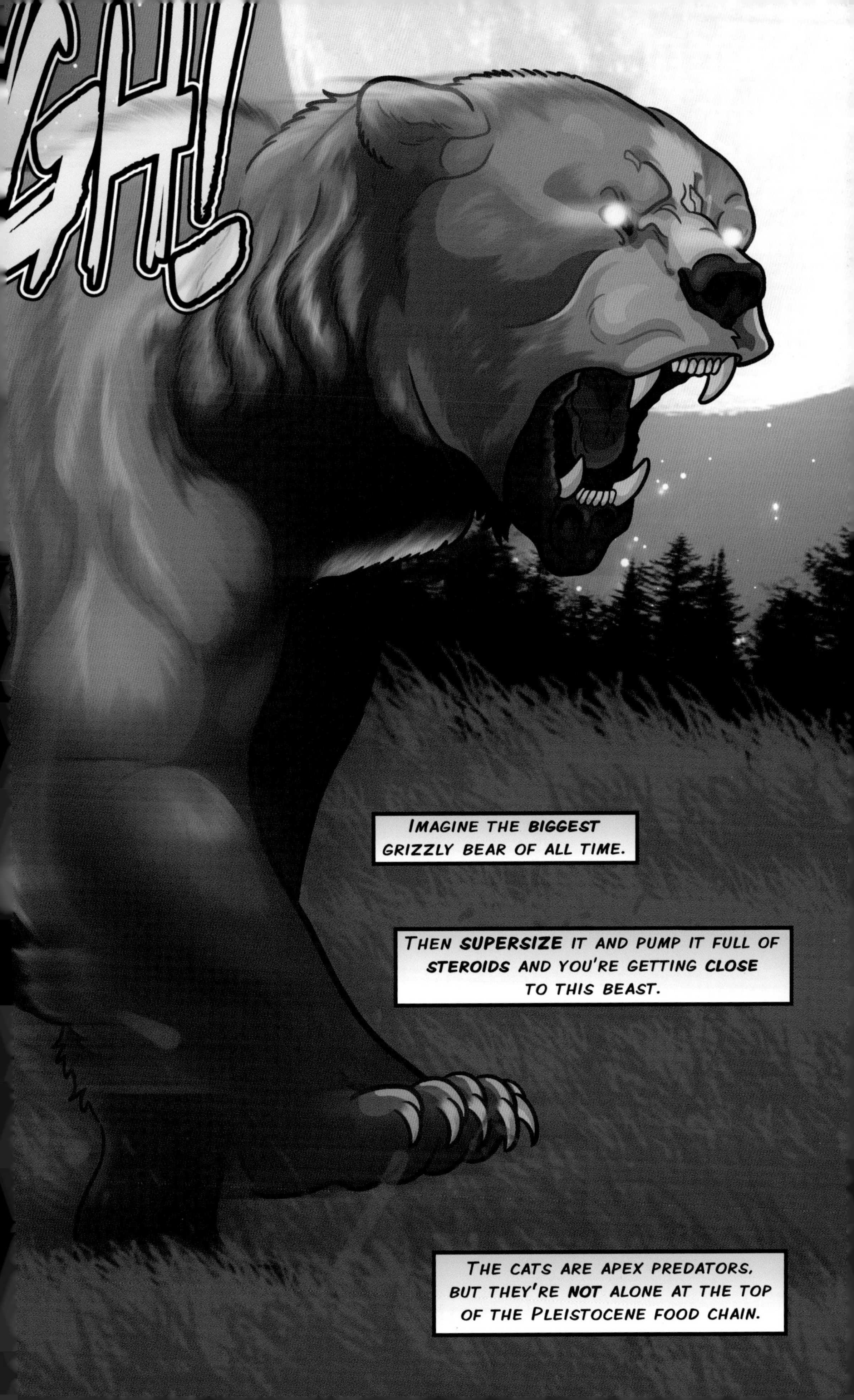
IMAGINE THE BIGGEST GRIZZLY BEAR OF ALL TIME.
THEN SUPERSIZE IT AND PUMP IT FULL OF STEROIDS AND YOU'RE GETTING CLOSE TO THIS BEAST.
THE CATS ARE APEX PREDATORS, BUT THEY'RE NOT ALONE AT THE TOP OF THE PLEISTOCENE FOOD CHAIN.

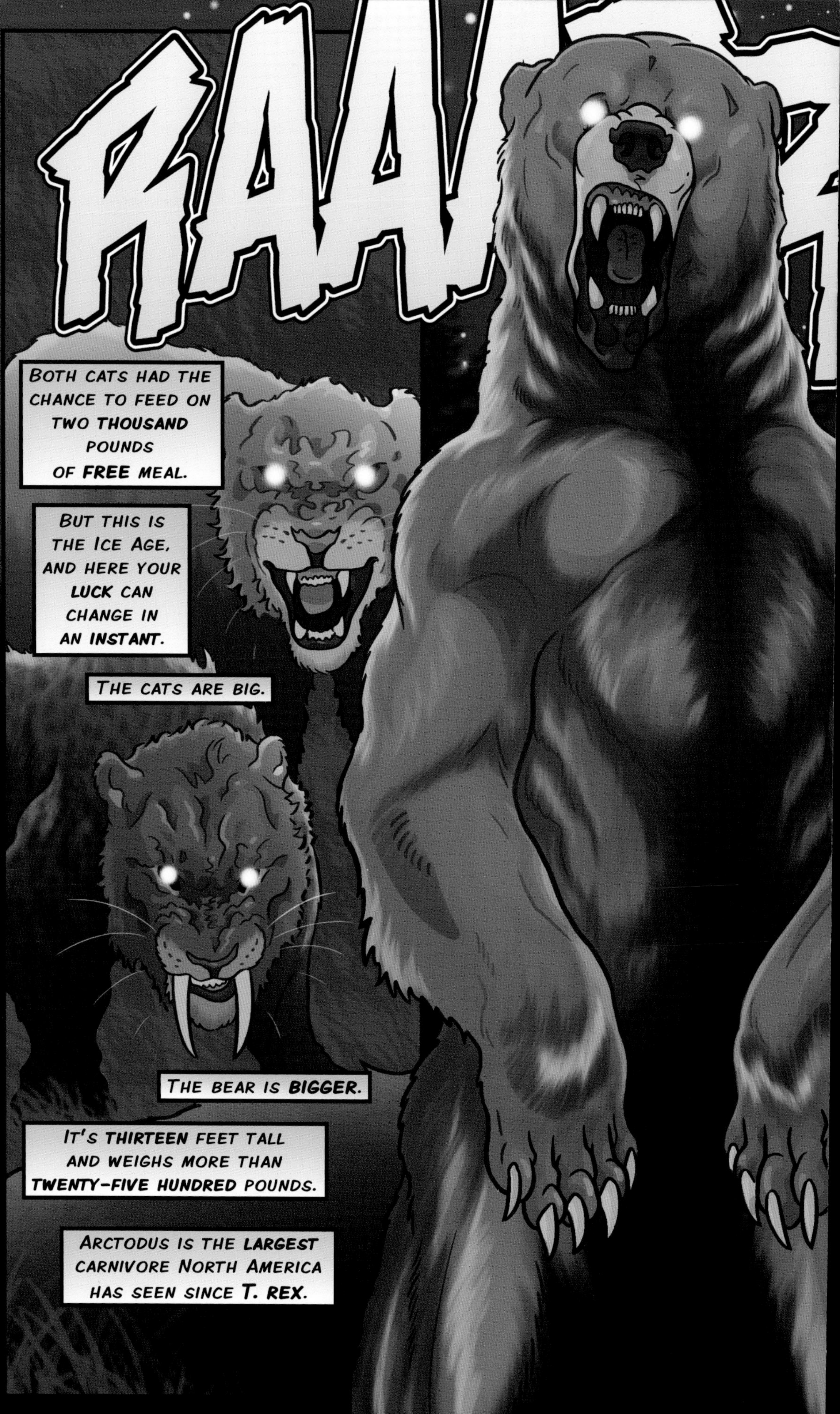
Both cats had the chance to feed on two thousand pounds of free meal.
But this is the Ice Age, and here your luck can change in an instant.
The cats are big.
The bear is bigger.
It's thirteen feet tall and weighs more than twenty-five hundred pounds.
Arctodus is the largest carnivore North America has seen since T. rex.

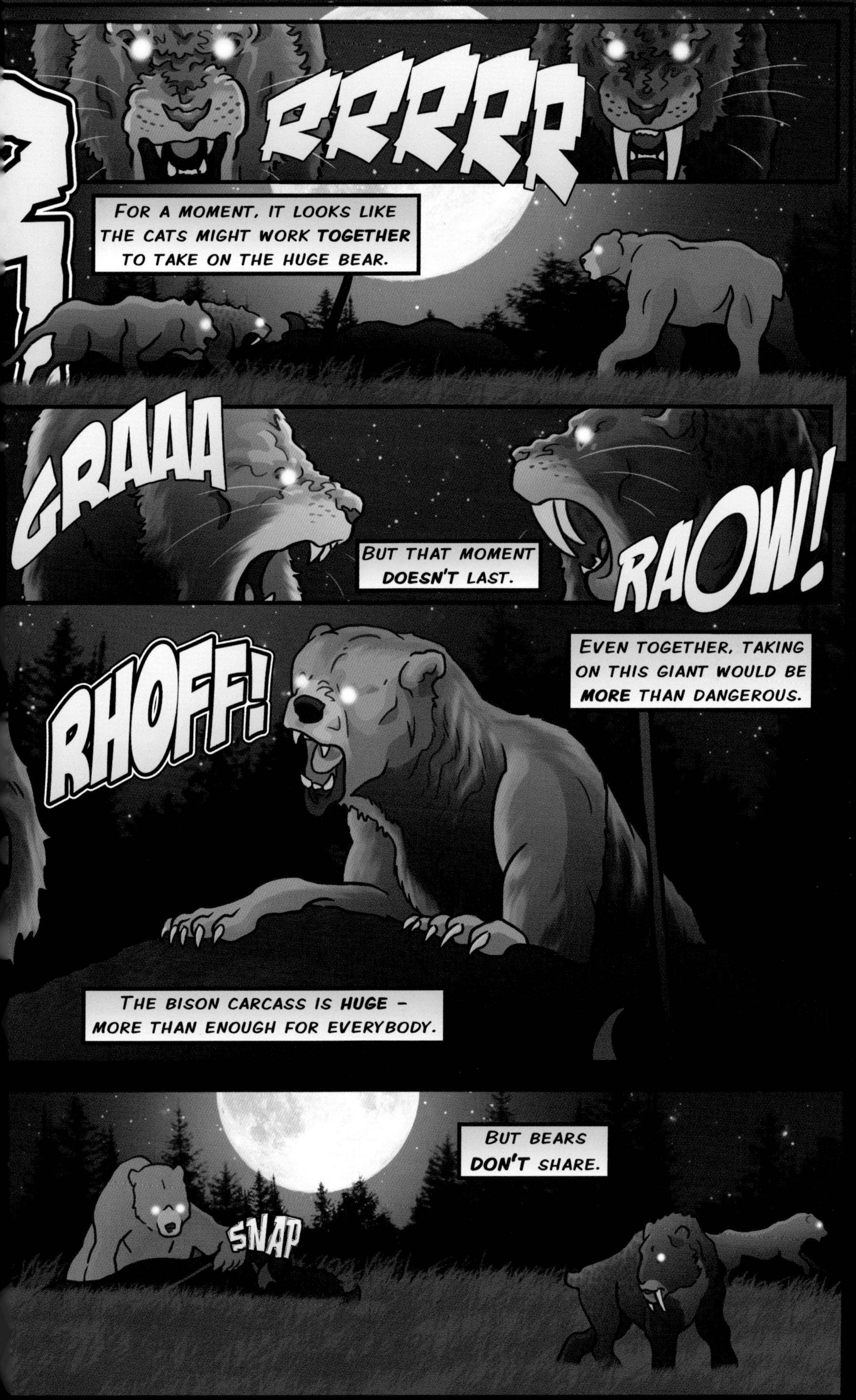
RRRRRR
FOR A MOMENT, IT LOOKS LIKE THE CATS MIGHT WORK TOGETHER TO TAKE ON THE HUGE BEAR.
GRAAA
BUT THAT MOMENT DOESN'T LAST.
RAOW!
RHOFF!
EVEN TOGETHER, TAKING ON THIS GIANT WOULD BE MORE THAN DANGEROUS.
THE BISON CARCASS IS HUGE – MORE THAN ENOUGH FOR EVERYBODY.
BUT BEARS DON'T SHARE.
SNAP

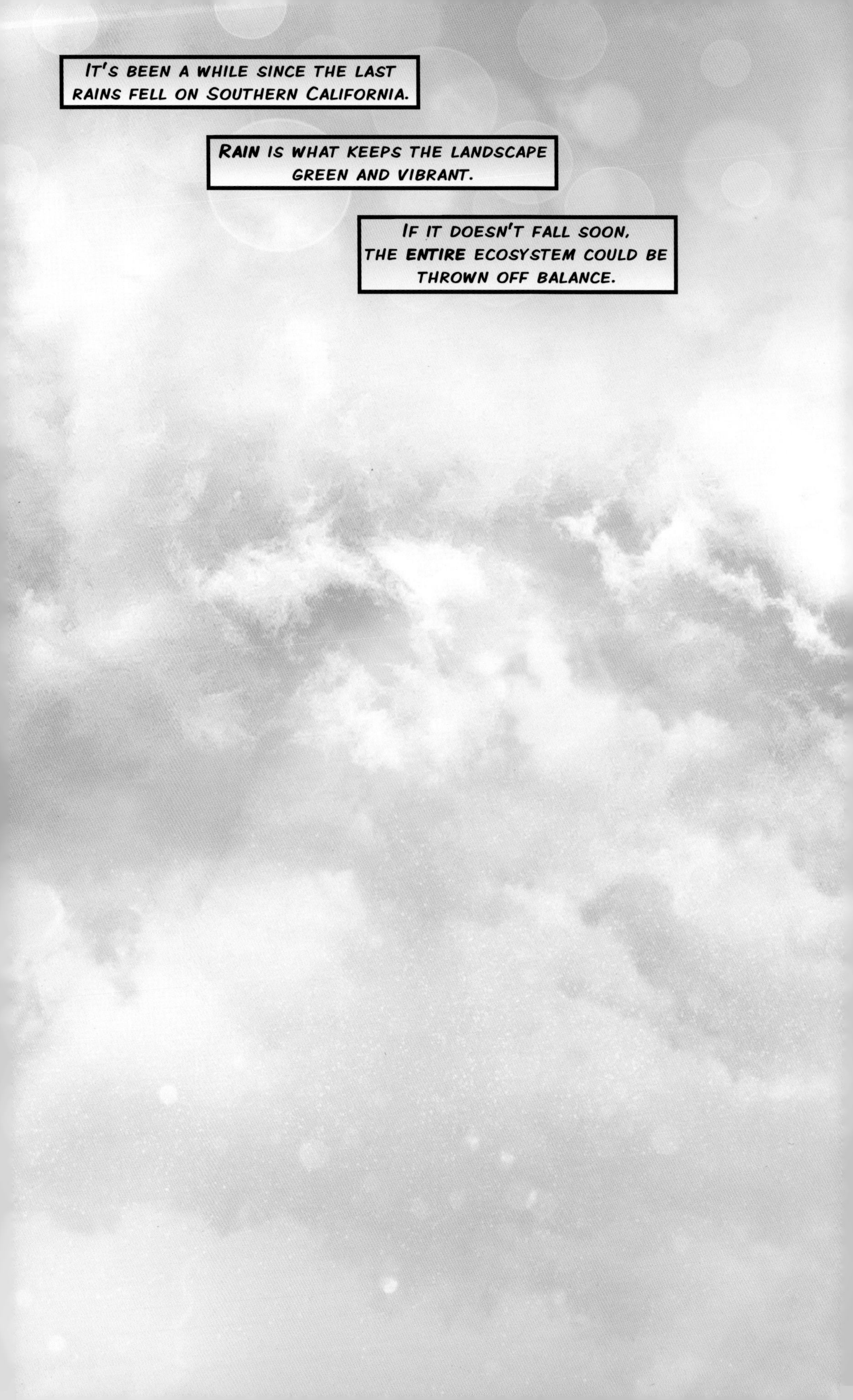
It's been a while since the last rains fell on Southern California.
Rain is what keeps the landscape green and vibrant.
If it doesn't fall soon, the entire ecosystem could be thrown off balance.

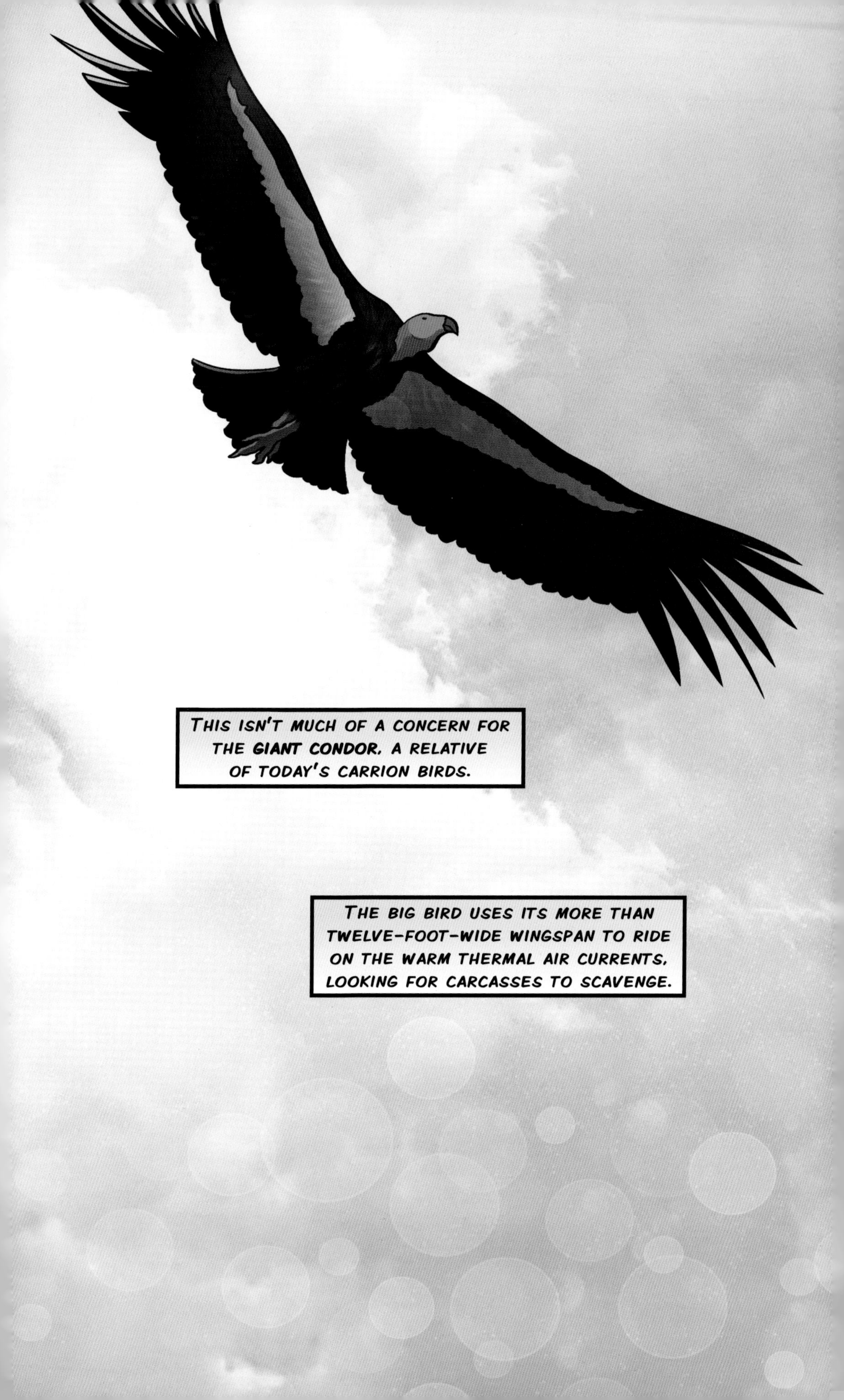
THIS ISN'T MUCH OF A CONCERN FOR THE **GIANT CONDOR**, A RELATIVE OF TODAY'S CARRION BIRDS.
THE BIG BIRD USES ITS MORE THAN TWELVE-FOOT-WIDE WINGSPAN TO RIDE ON THE WARM THERMAL AIR CURRENTS, LOOKING FOR CARCASSES TO SCAVENGE.

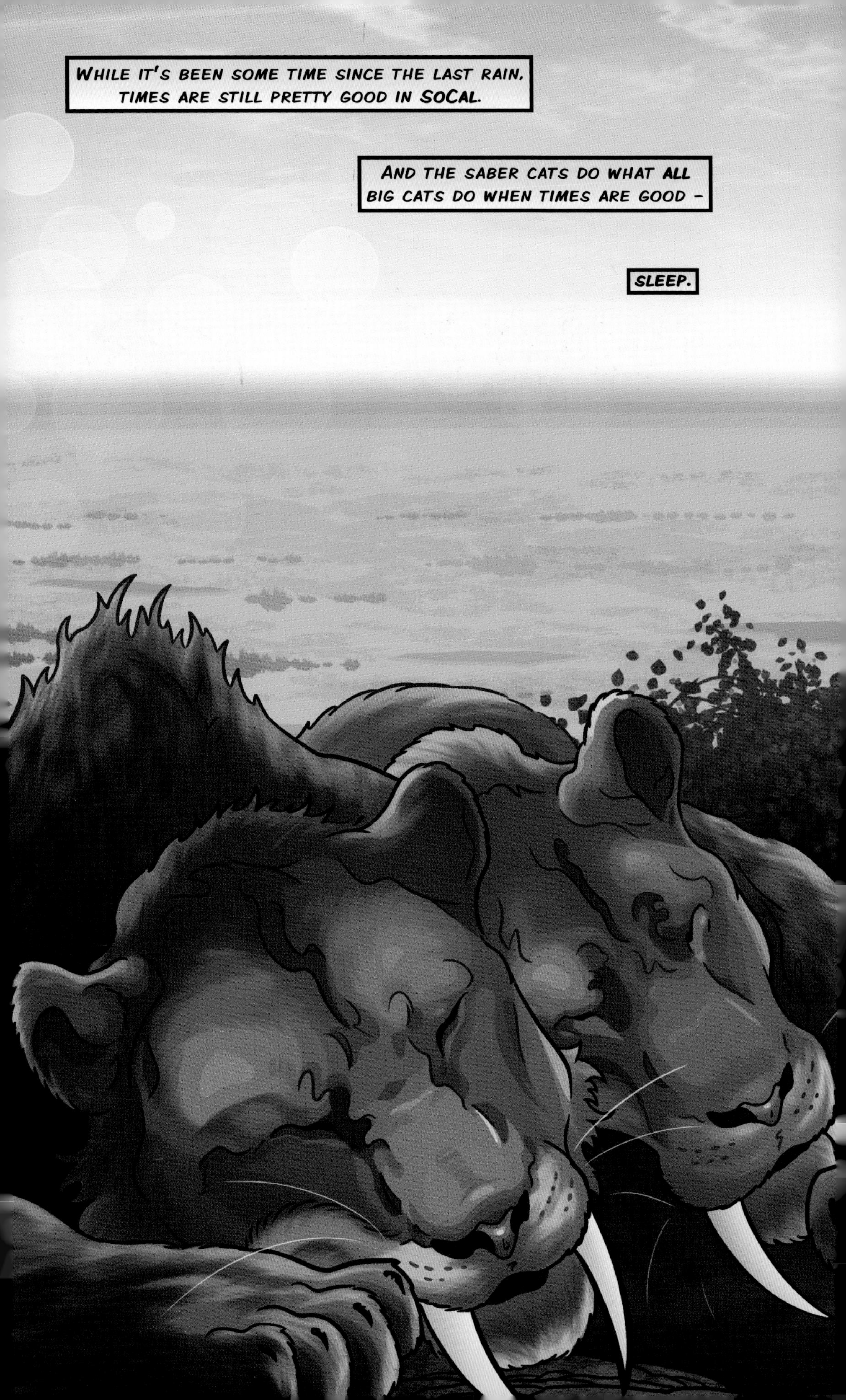
WHILE IT'S BEEN SOME TIME SINCE THE LAST RAIN, TIMES ARE STILL PRETTY GOOD IN SOCAL.
AND THE SABER CATS DO WHAT ALL BIG CATS DO WHEN TIMES ARE GOOD -
SLEEP.

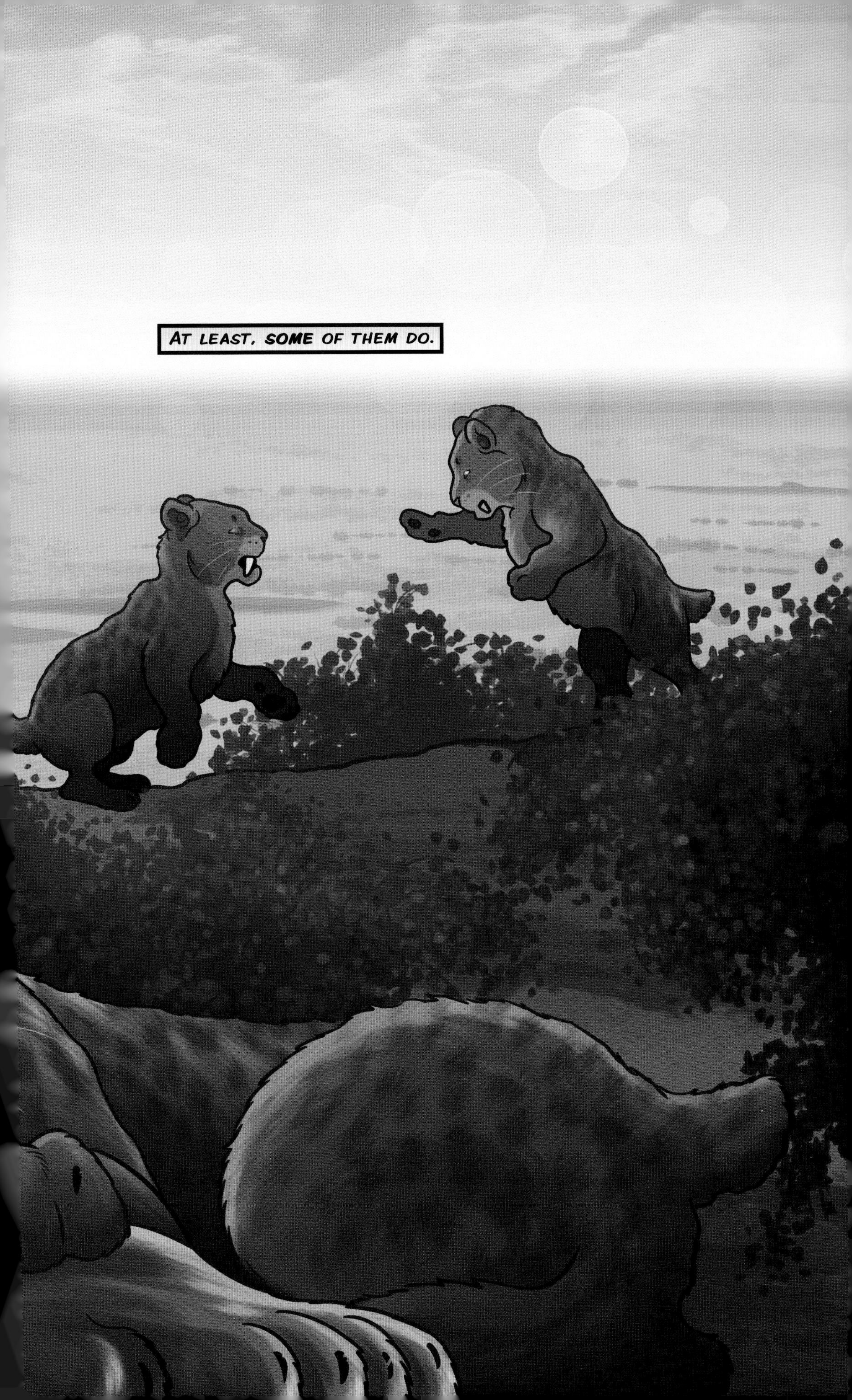
AT LEAST, SOME OF THEM DO.

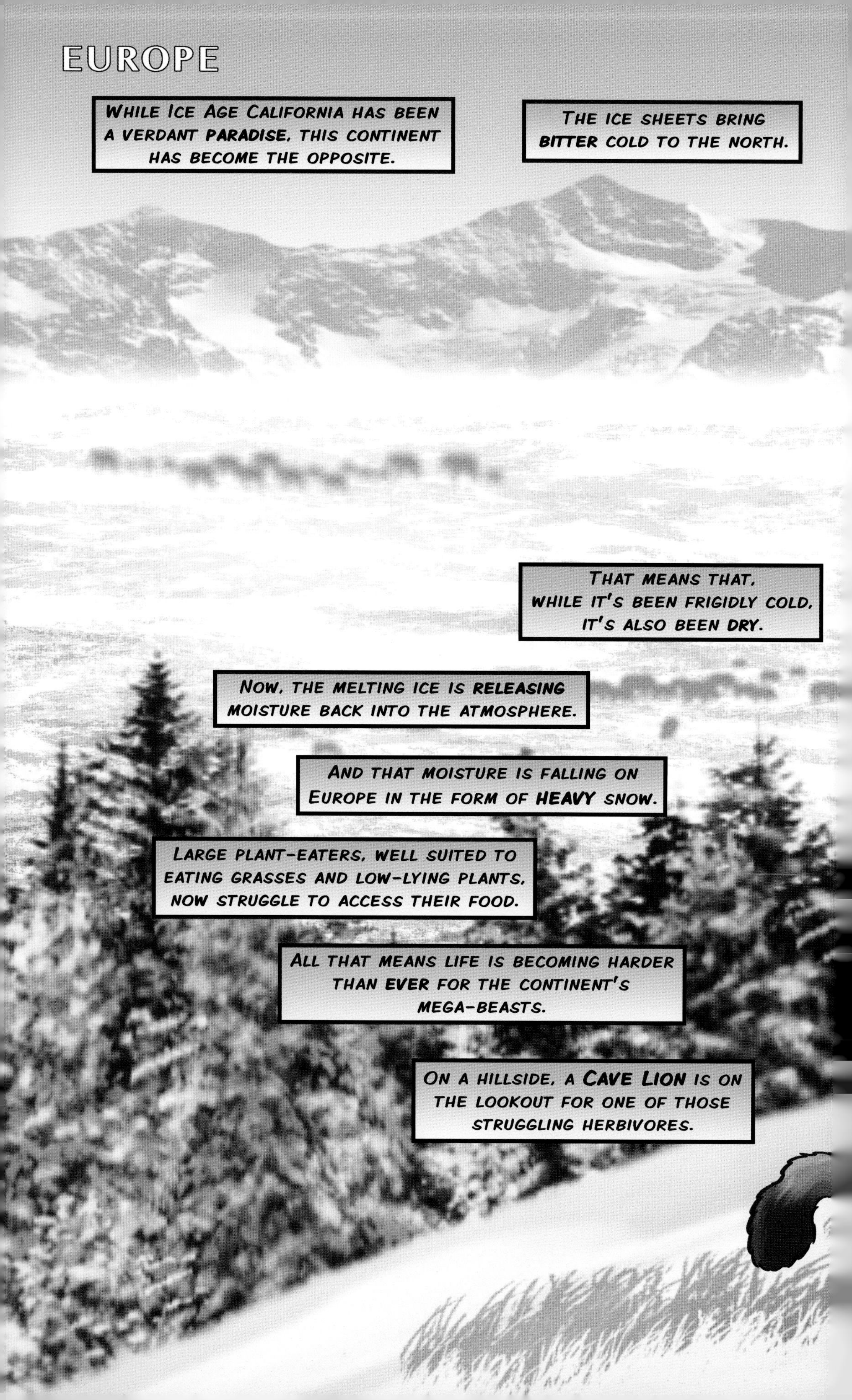
EUROPE
WHILE ICE AGE CALIFORNIA HAS BEEN A VERDANT PARADISE, THIS CONTINENT HAS BECOME THE OPPOSITE.
THE ICE SHEETS BRING BITTER COLD TO THE NORTH.
THAT MEANS THAT, WHILE IT'S BEEN FRIGIDLY COLD, IT'S ALSO BEEN DRY.
NOW, THE MELTING ICE IS RELEASING MOISTURE BACK INTO THE ATMOSPHERE.
AND THAT MOISTURE IS FALLING ON EUROPE IN THE FORM OF HEAVY SNOW.
LARGE PLANT-EATERS, WELL SUITED TO EATING GRASSES AND LOW-LYING PLANTS, NOW STRUGGLE TO ACCESS THEIR FOOD.
ALL THAT MEANS LIFE IS BECOMING HARDER THAN EVER FOR THE CONTINENT'S MEGA-BEASTS.
ON A HILLSIDE, A CAVE LION IS ON THE LOOKOUT FOR ONE OF THOSE STRUGGLING HERBIVORES.

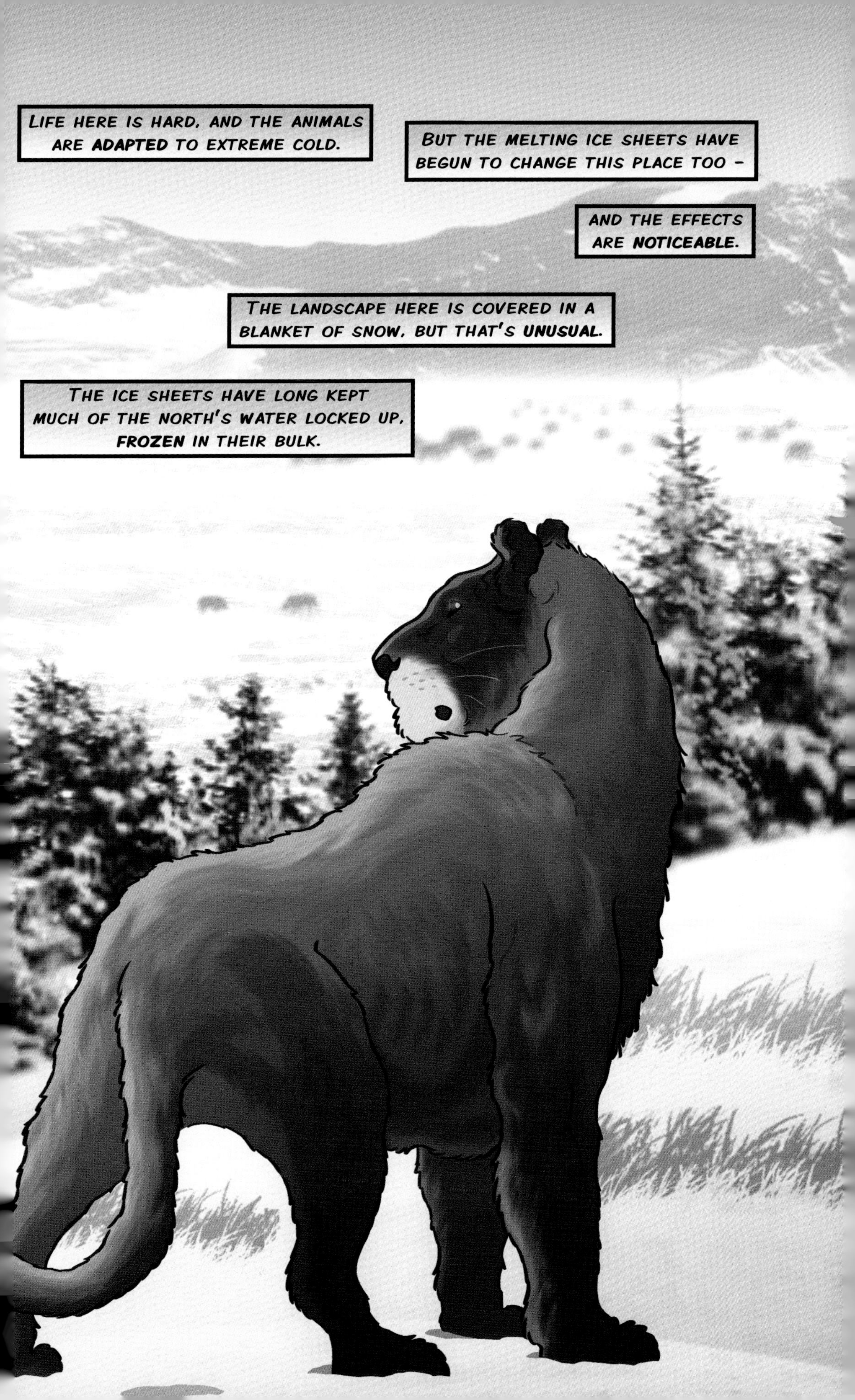
LIFE HERE IS HARD, AND THE ANIMALS ARE ADAPTED TO EXTREME COLD.
BUT THE MELTING ICE SHEETS HAVE BEGUN TO CHANGE THIS PLACE TOO –
AND THE EFFECTS ARE NOTICEABLE.
THE LANDSCAPE HERE IS COVERED IN A BLANKET OF SNOW, BUT THAT'S UNUSUAL.
THE ICE SHEETS HAVE LONG KEPT MUCH OF THE NORTH'S WATER LOCKED UP, FROZEN IN THEIR BULK.

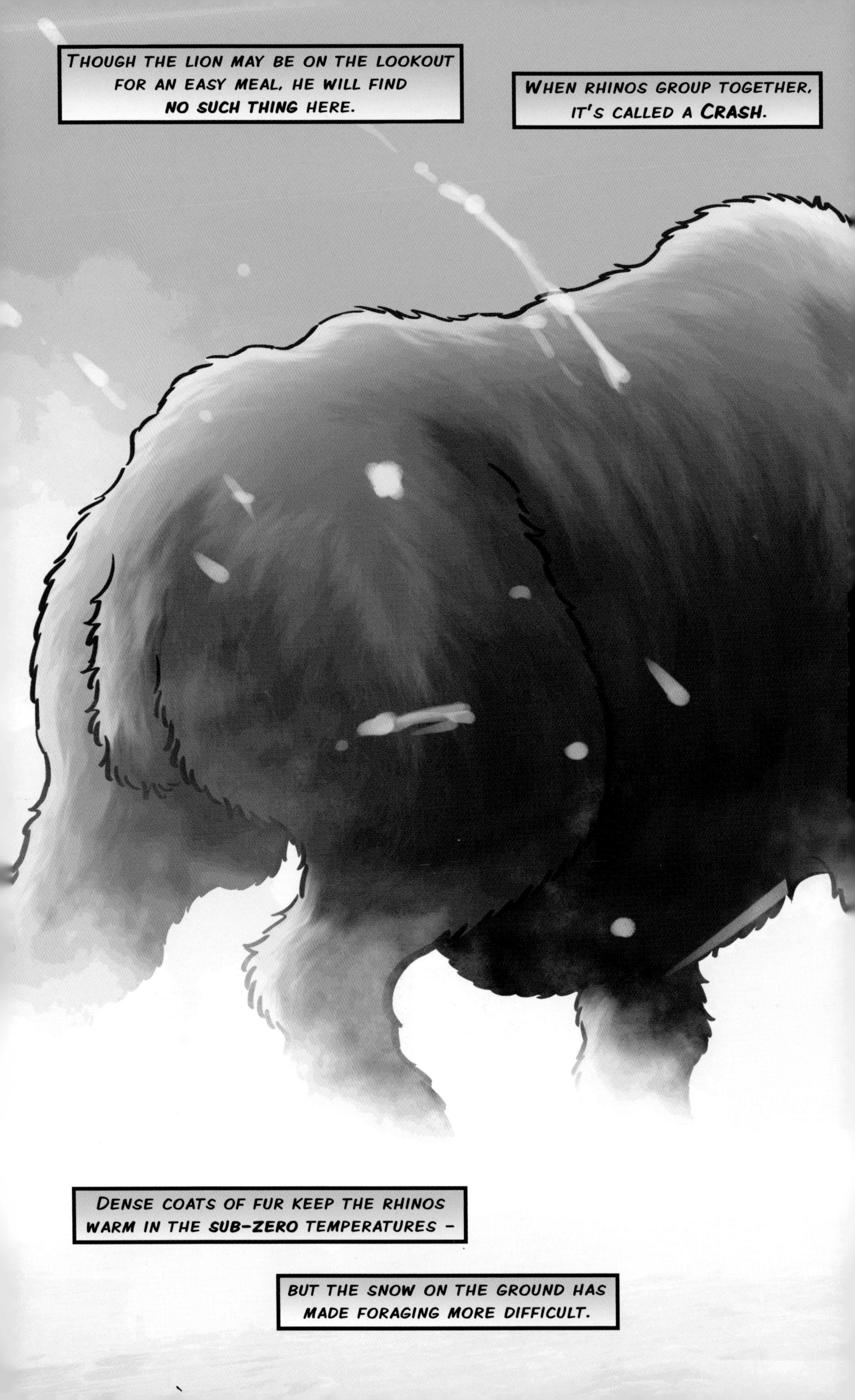
THOUGH THE LION MAY BE ON THE LOOKOUT FOR AN EASY MEAL, HE WILL FIND **NO SUCH THING** HERE.
WHEN RHINOS GROUP TOGETHER, IT'S CALLED A **CRASH**.
DENSE COATS OF FUR KEEP THE RHINOS WARM IN THE **SUB-ZERO** TEMPERATURES –
BUT THE SNOW ON THE GROUND HAS MADE FORAGING MORE DIFFICULT.

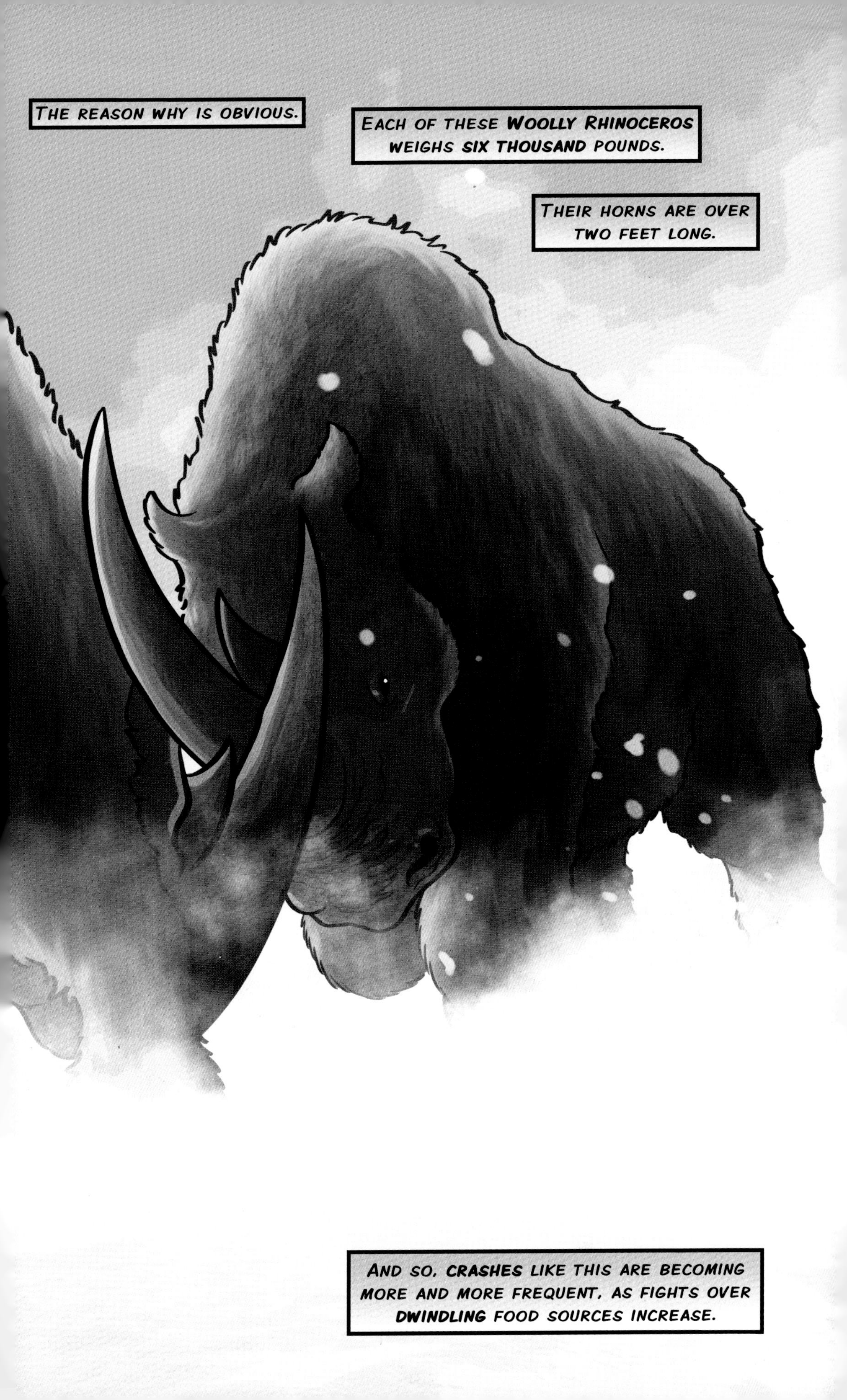
THE REASON WHY IS OBVIOUS.
EACH OF THESE WOOLLY RHINOCEROS WEIGHS SIX THOUSAND POUNDS.
THEIR HORNS ARE OVER TWO FEET LONG.
AND SO, CRASHES LIKE THIS ARE BECOMING MORE AND MORE FREQUENT, AS FIGHTS OVER DWINDLING FOOD SOURCES INCREASE.

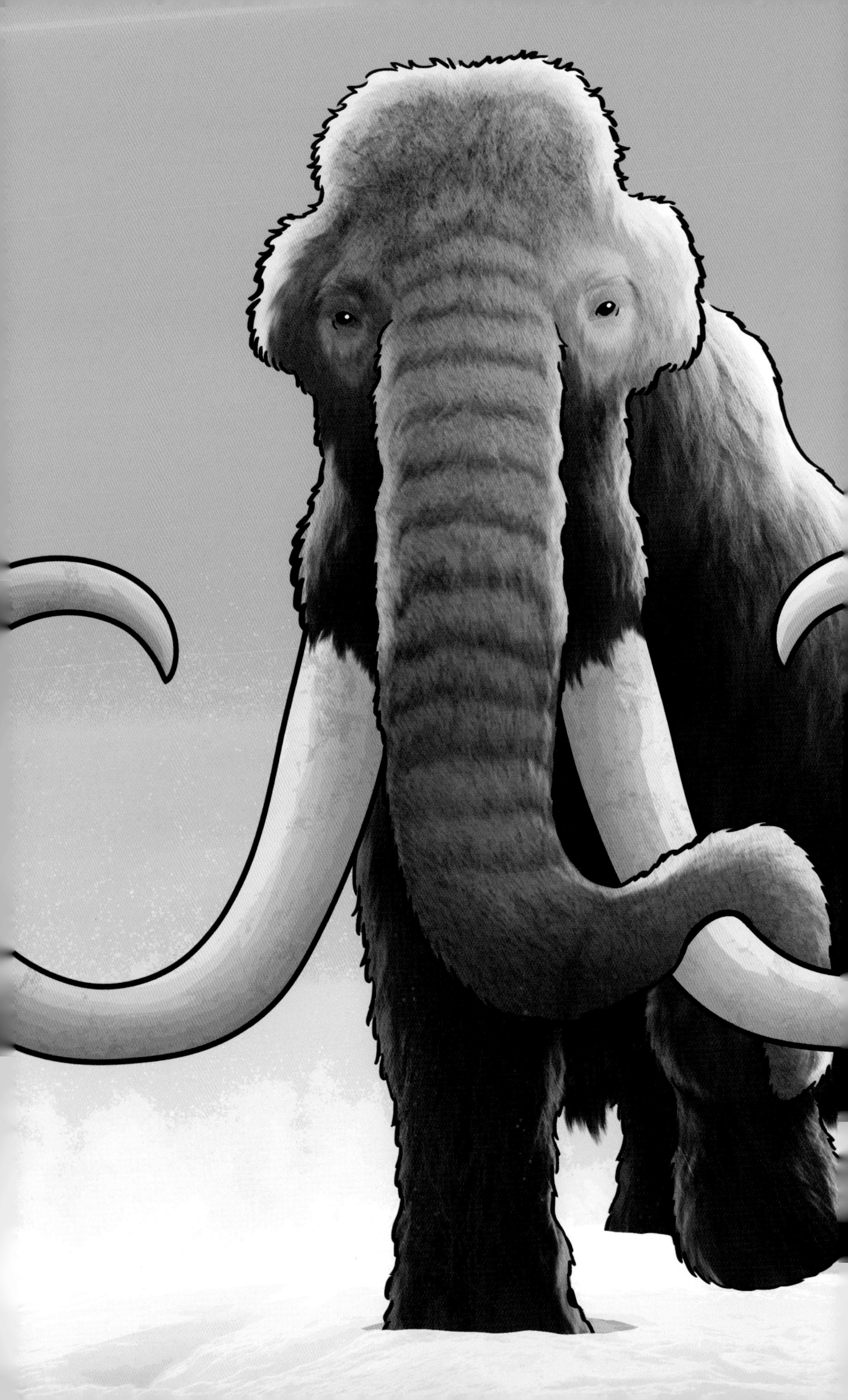

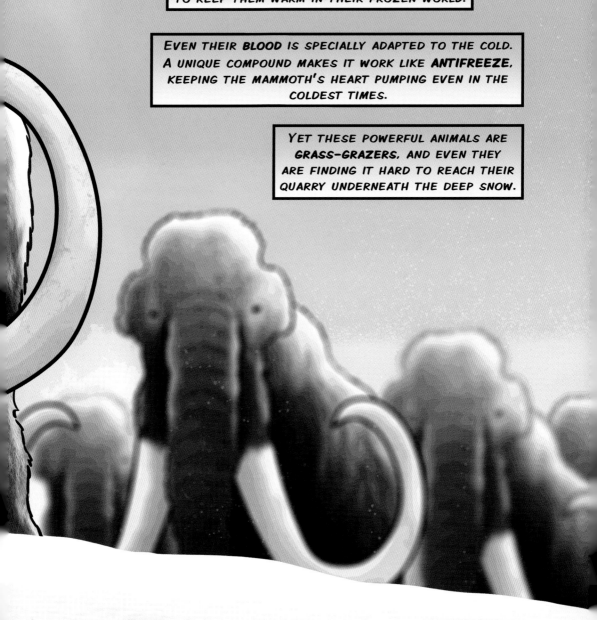

THIS PLACE IS KNOWN AS THE **MAMMOTH STEPPE,** AND THESE SHAGGY BEASTS ARE THE REASON WHY.

WOOLLY MAMMOTHS ARE PERHAPS THE MOST WELL KNOWN OF ALL THE PLEISTOCENE'S CHARISMATIC MEGAFAUNA.

THESE HIRSUTE ELEPHANTS ARE THE MOST **SPECIALIZED** OF THEIR GROUP.

THEY'RE TEN FEET TALL AND WEIGH MORE THAN FIVE TONS.

THICK FAT RESERVES AND A HEAVY COAT HELP TO KEEP THEM WARM IN THEIR FROZEN WORLD.

EVEN THEIR **BLOOD** IS SPECIALLY ADAPTED TO THE COLD. A UNIQUE COMPOUND MAKES IT WORK LIKE **ANTIFREEZE,** KEEPING THE MAMMOTH'S HEART PUMPING EVEN IN THE COLDEST TIMES.

YET THESE POWERFUL ANIMALS ARE **GRASS-GRAZERS,** AND EVEN THEY ARE FINDING IT HARD TO REACH THEIR QUARRY UNDERNEATH THE DEEP SNOW.

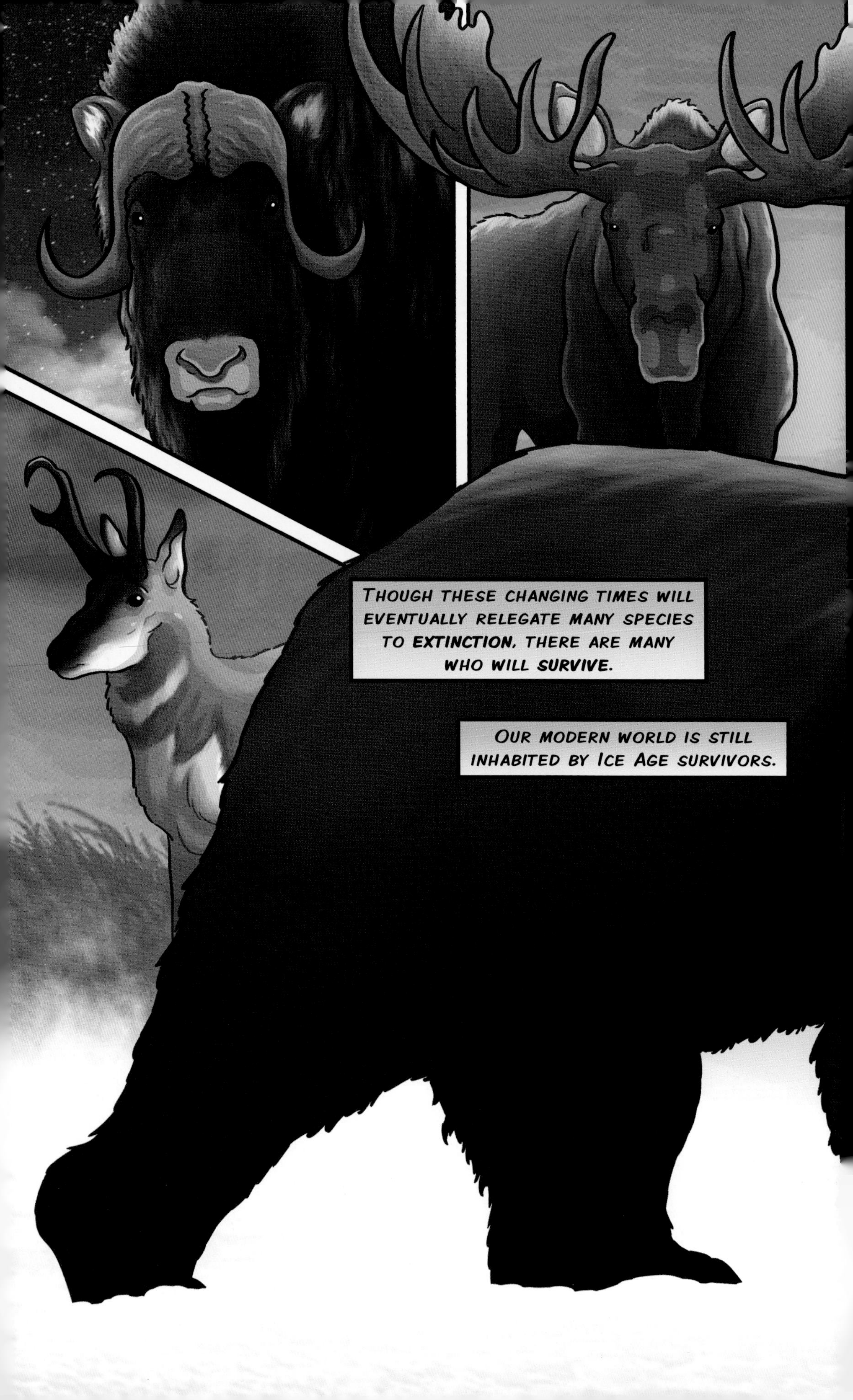
THOUGH THESE CHANGING TIMES WILL EVENTUALLY RELEGATE MANY SPECIES TO EXTINCTION, THERE ARE MANY WHO WILL SURVIVE.
OUR MODERN WORLD IS STILL INHABITED BY ICE AGE SURVIVORS.

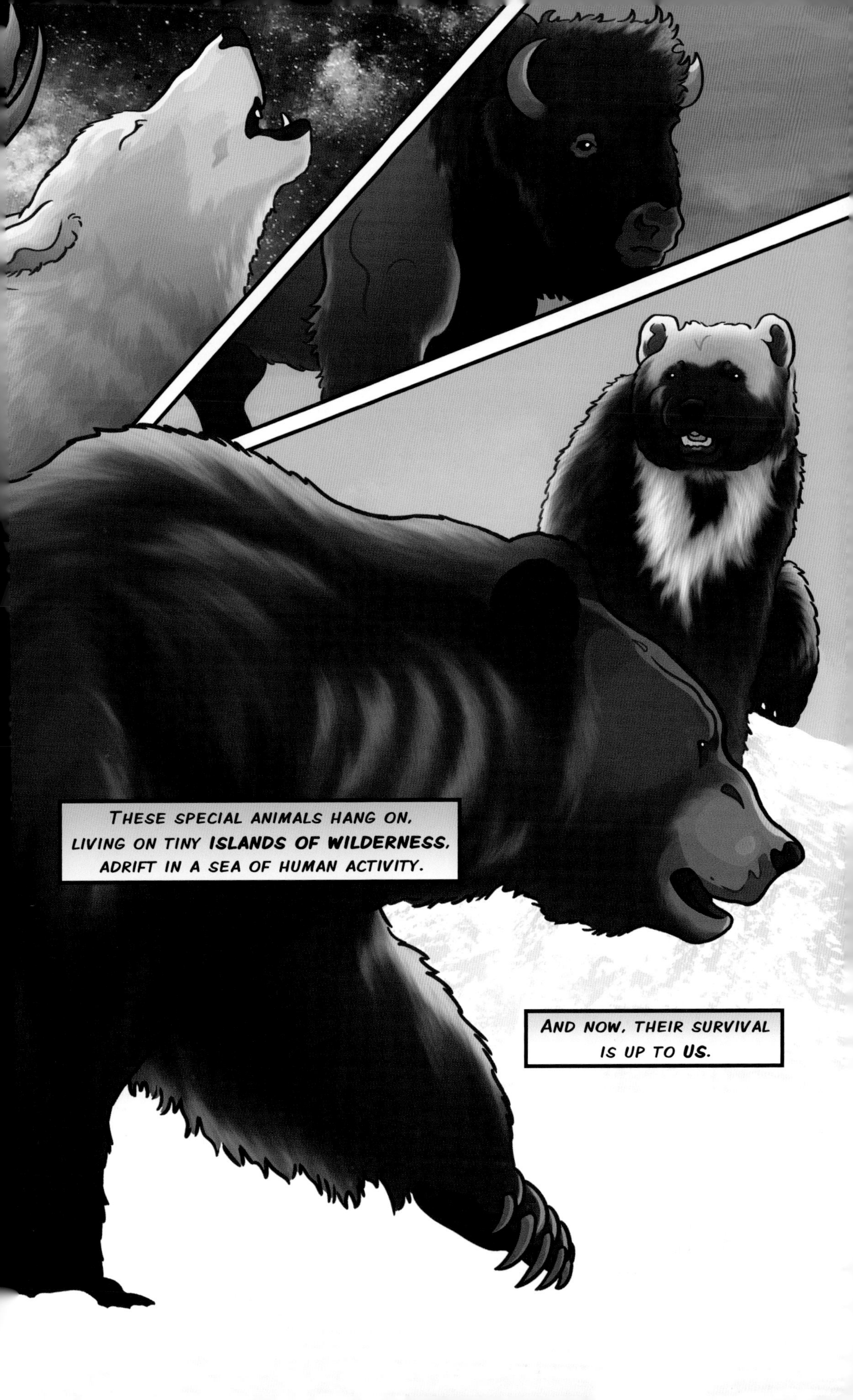
THESE SPECIAL ANIMALS HANG ON, LIVING ON TINY ISLANDS OF WILDERNESS, ADRIFT IN A SEA OF HUMAN ACTIVITY.
AND NOW, THEIR SURVIVAL IS UP TO US.

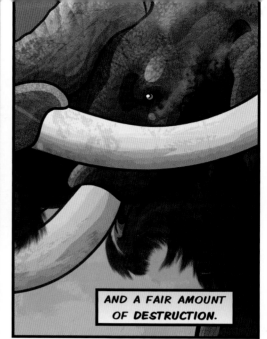

BACK IN SOUTHERN CALIFORNIA, TWO MALE COLUMBIAN MAMMOTHS ARE ENGAGING IN A LITTLE BIT OF CHAOS –

AND A FAIR AMOUNT OF DESTRUCTION.

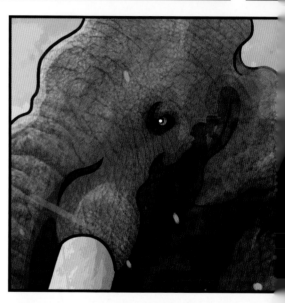

THEY'RE READY TO MATE, AND TESTOSTERONE IS SURGING THROUGH THEIR BLOODSTREAM, TELLING THEM TO PROVE THEIR WORTH TO FEMALE MAMMOTHS –

TELLING THEM TO SMASH.

THE NORMALLY GENTLE GIANTS HAVE BEEN TRANSFORMED INTO LIVING FREIGHT TRAINS OF AGGRESSION.

THIS IS CALLED **MUSTH**.

THE BULLS' BODIES ARE BEING **FLOODED** WITH HORMONES.

THIS ISN'T LIKE RUTTING DEER OR BISON. MAMMOTHS – LIKE **ALL** ELEPHANTS – DON'T NECESSARILY BREED AT THE SAME TIME EVERY YEAR.

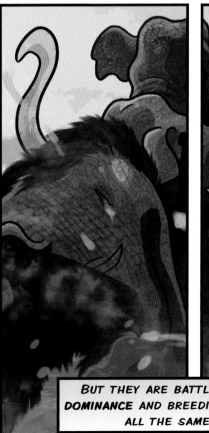

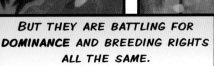

BUT THEY ARE BATTLING FOR **DOMINANCE** AND BREEDING RIGHTS ALL THE SAME.

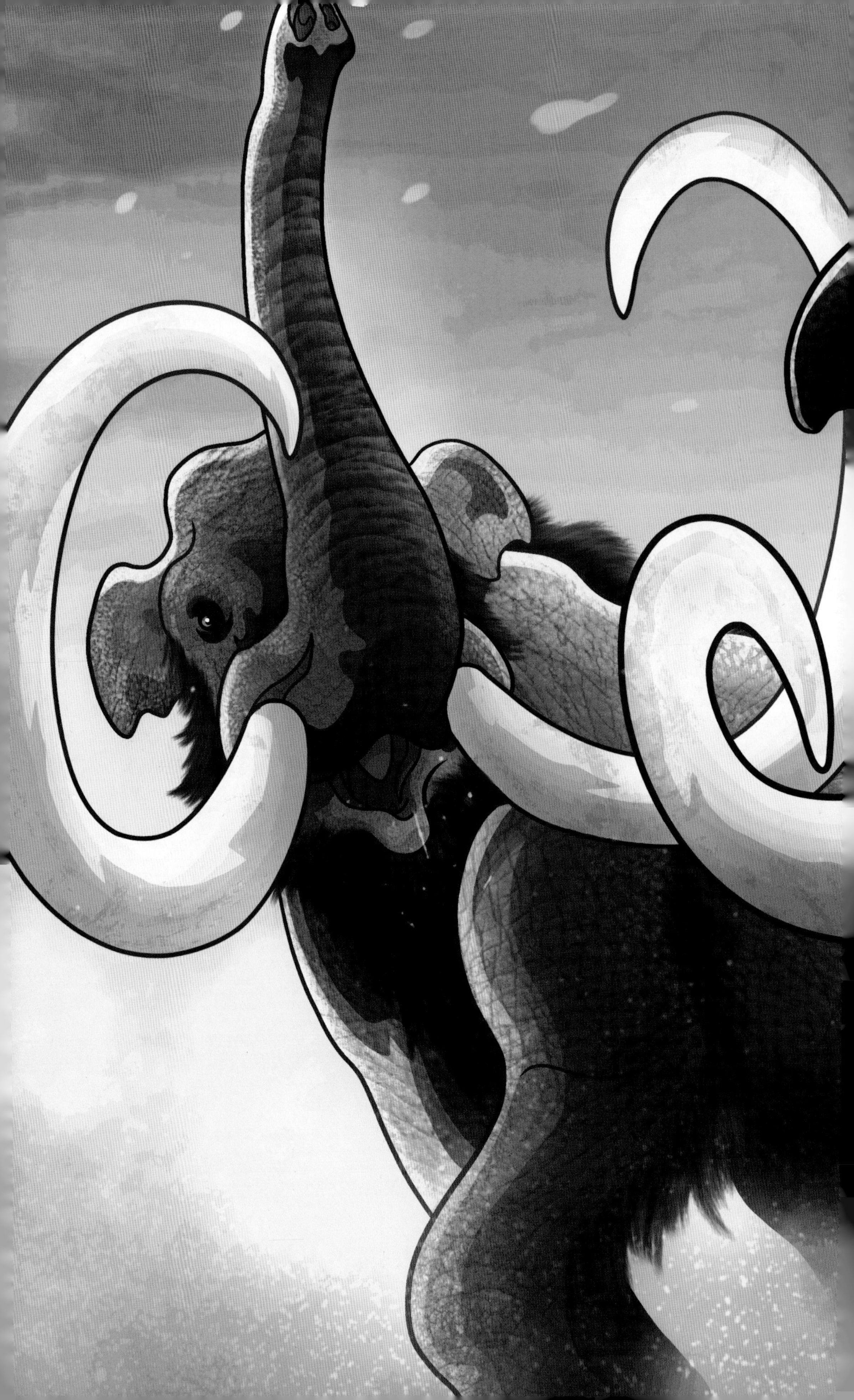

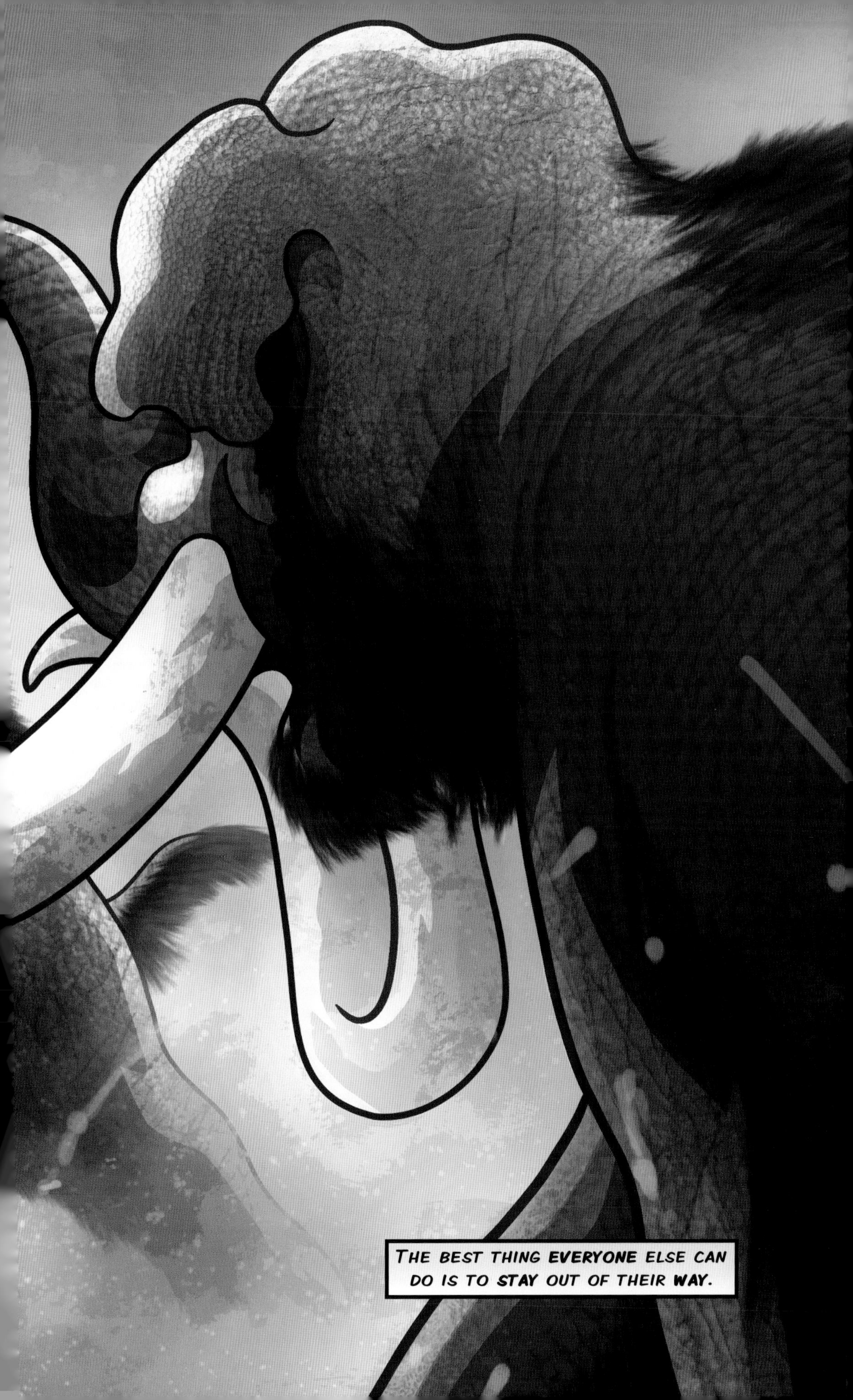
THE BEST THING EVERYONE ELSE CAN DO IS TO STAY OUT OF THEIR WAY.

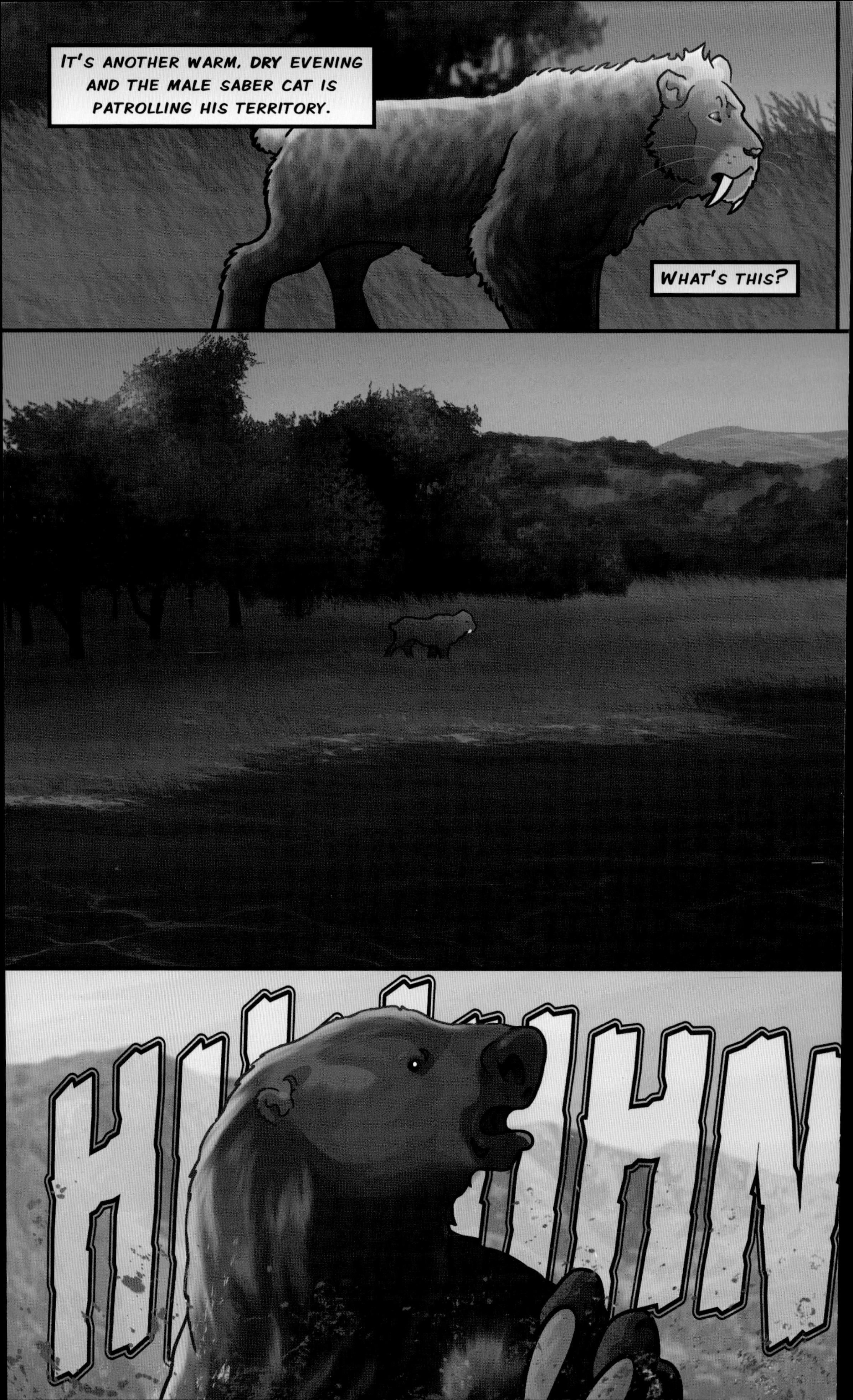
It's another warm, dry evening and the male saber cat is patrolling his territory.
What's this?

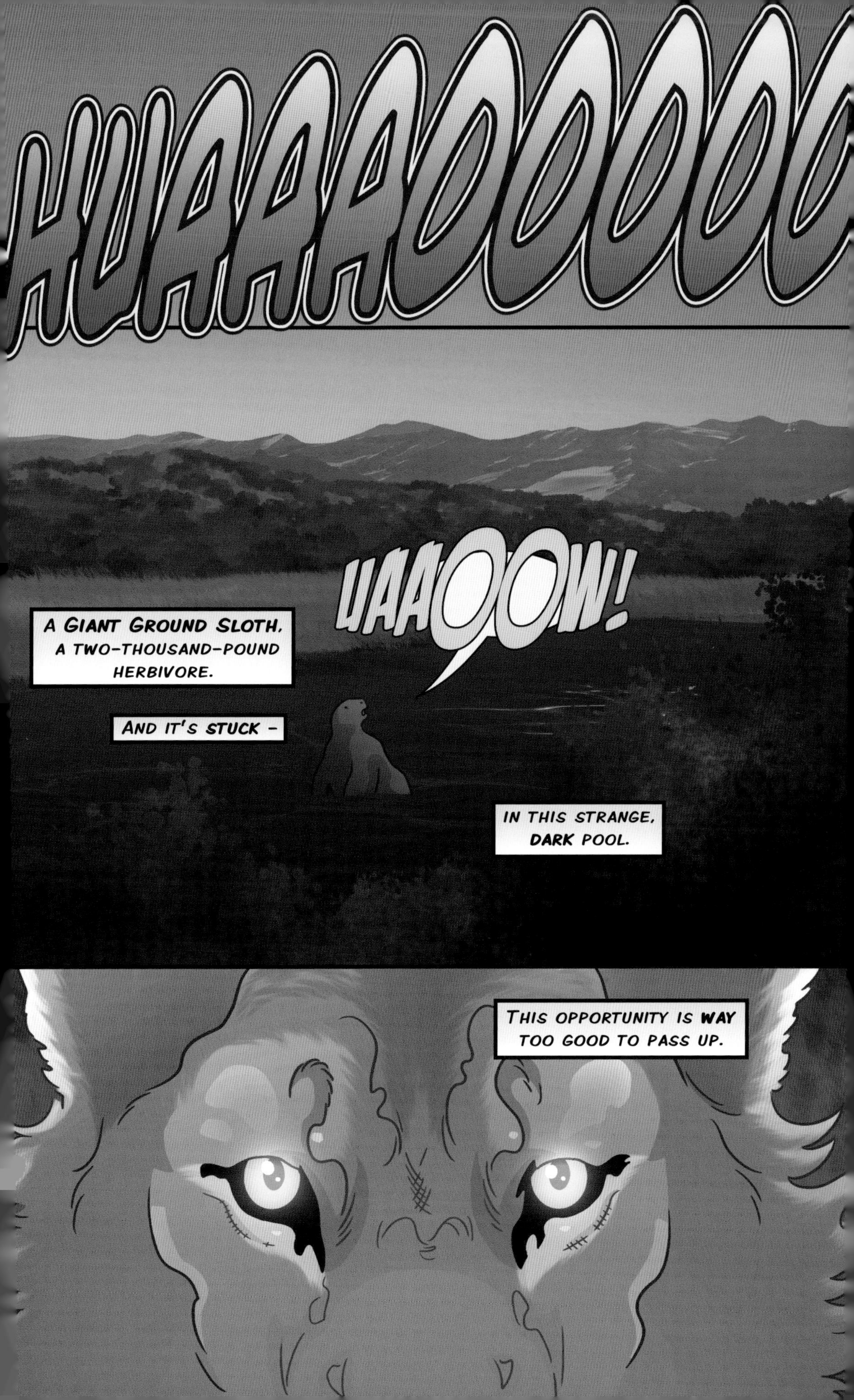
HUAAAOOOOO
A GIANT GROUND SLOTH, A TWO-THOUSAND-POUND HERBIVORE.
UAAOOOW!
AND IT'S STUCK –
IN THIS STRANGE, DARK POOL.
THIS OPPORTUNITY IS WAY TOO GOOD TO PASS UP.

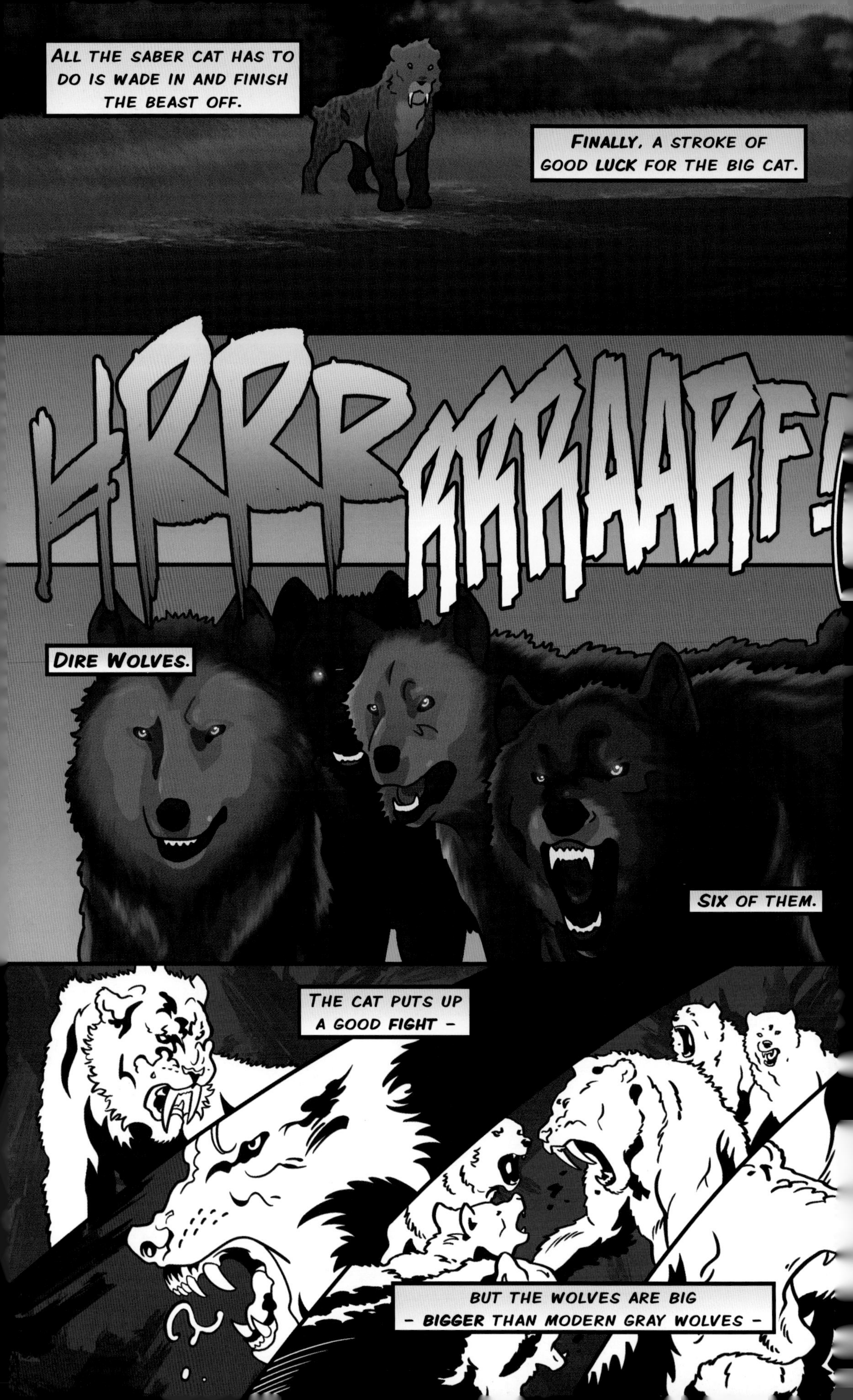
ALL THE SABER CAT HAS TO DO IS WADE IN AND FINISH THE BEAST OFF.
FINALLY, A STROKE OF GOOD LUCK FOR THE BIG CAT.
HRRRRRRAARF!
DIRE WOLVES.
SIX OF THEM.
THE CAT PUTS UP A GOOD FIGHT –
BUT THE WOLVES ARE BIG – BIGGER THAN MODERN GRAY WOLVES –

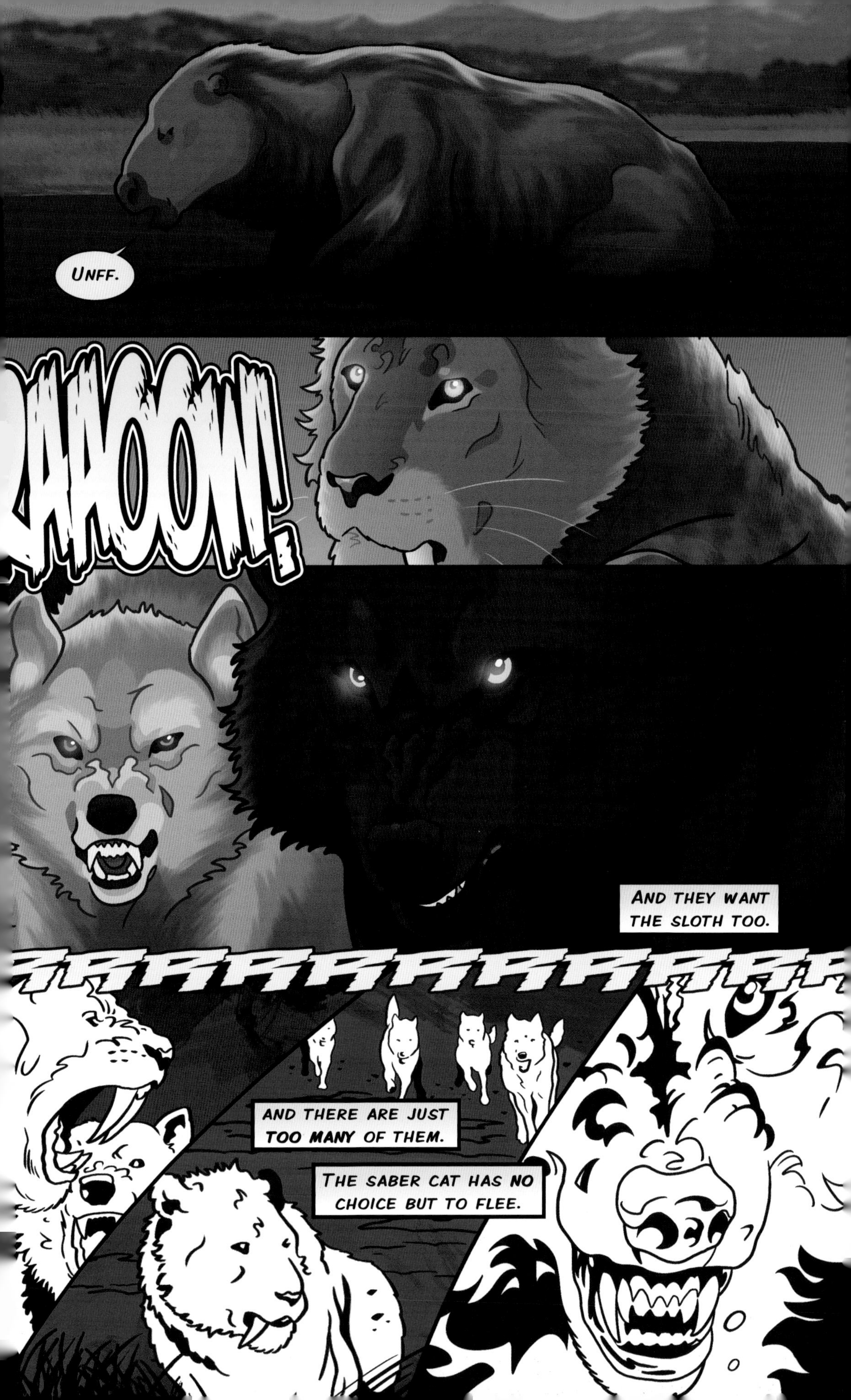
UNFF.
RAOOW!
AND THEY WANT THE SLOTH TOO.
RRRRRRRRRRRRR
AND THERE ARE JUST TOO MANY OF THEM.
THE SABER CAT HAS NO CHOICE BUT TO FLEE.

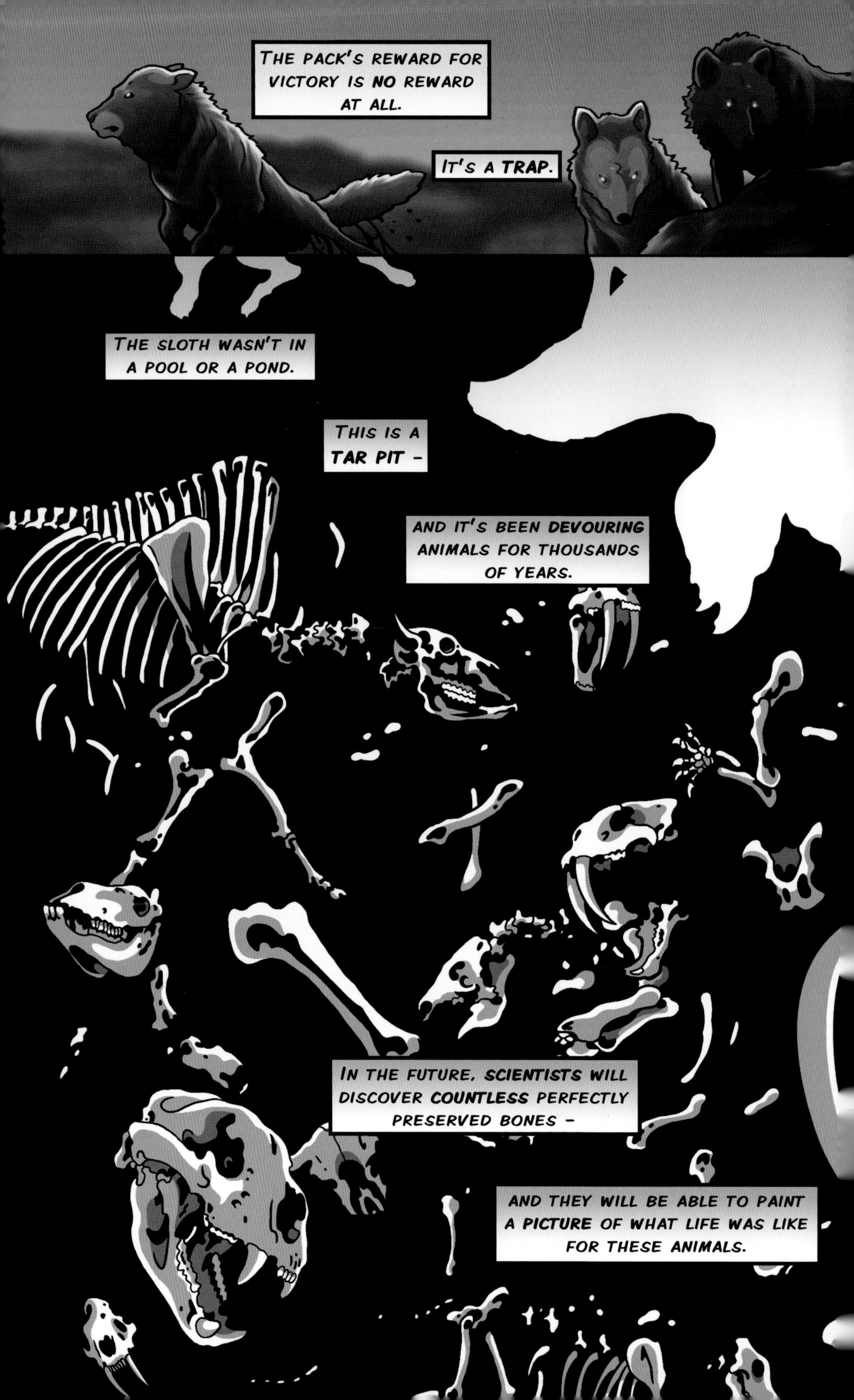
THE PACK'S REWARD FOR VICTORY IS NO REWARD AT ALL.
IT'S A TRAP.
THE SLOTH WASN'T IN A POOL OR A POND.
THIS IS A TAR PIT -
AND IT'S BEEN DEVOURING ANIMALS FOR THOUSANDS OF YEARS.
IN THE FUTURE, SCIENTISTS WILL DISCOVER COUNTLESS PERFECTLY PRESERVED BONES -
AND THEY WILL BE ABLE TO PAINT A PICTURE OF WHAT LIFE WAS LIKE FOR THESE ANIMALS.

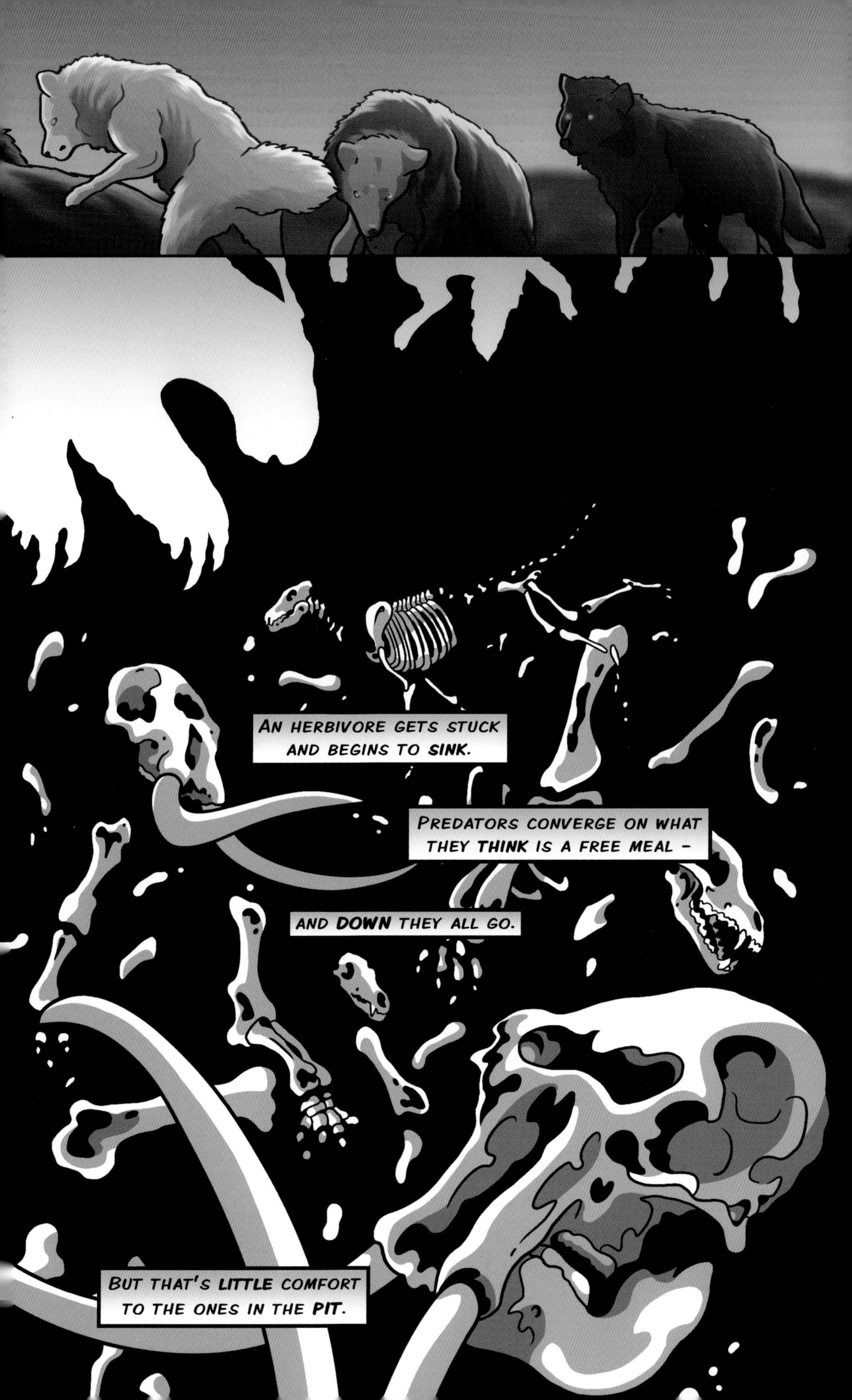
AN HERBIVORE GETS STUCK AND BEGINS TO SINK.
PREDATORS CONVERGE ON WHAT THEY THINK IS A FREE MEAL -
AND DOWN THEY ALL GO.
BUT THAT'S LITTLE COMFORT TO THE ONES IN THE PIT.

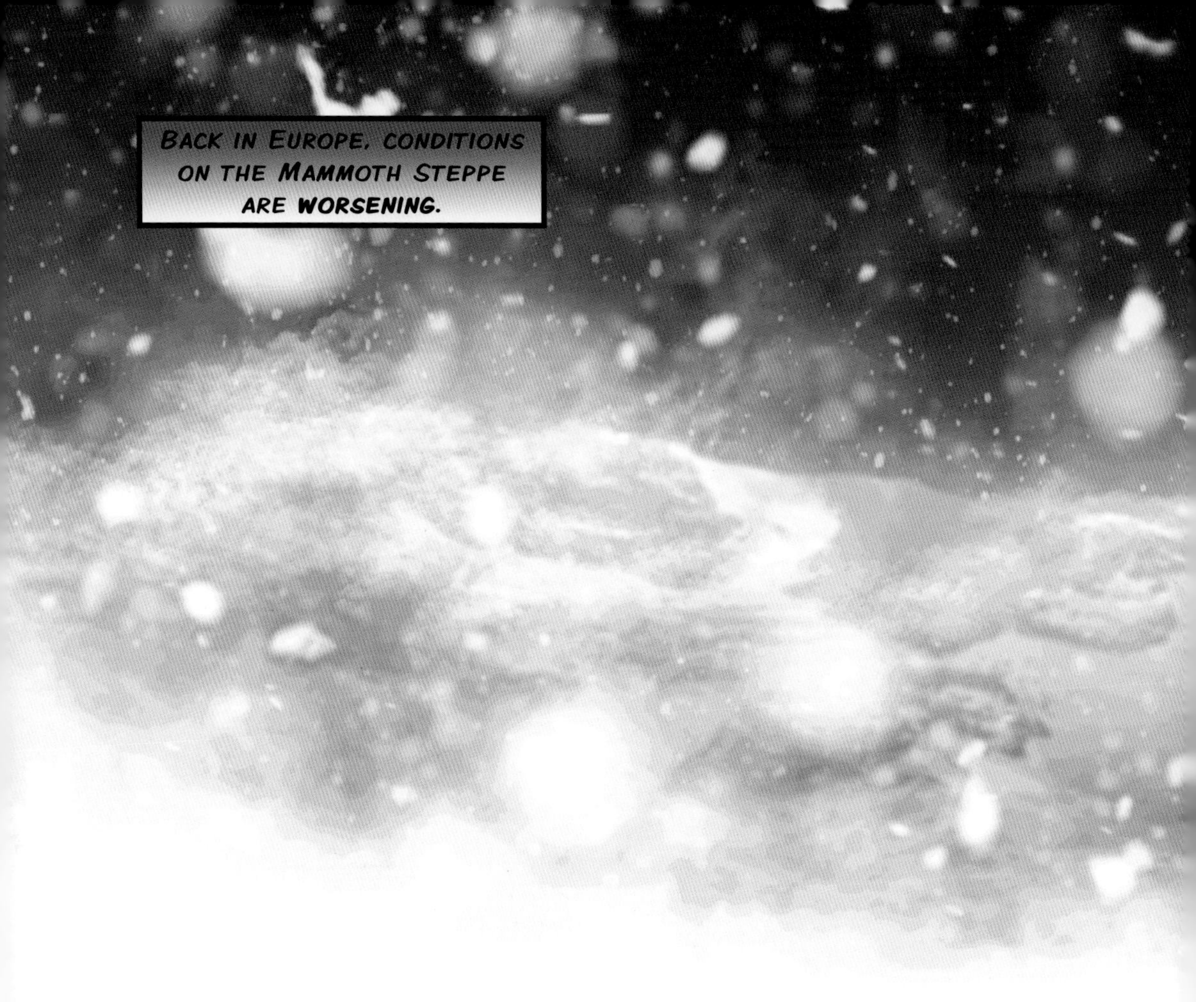
Back in Europe, conditions on the Mammoth Steppe are worsening.

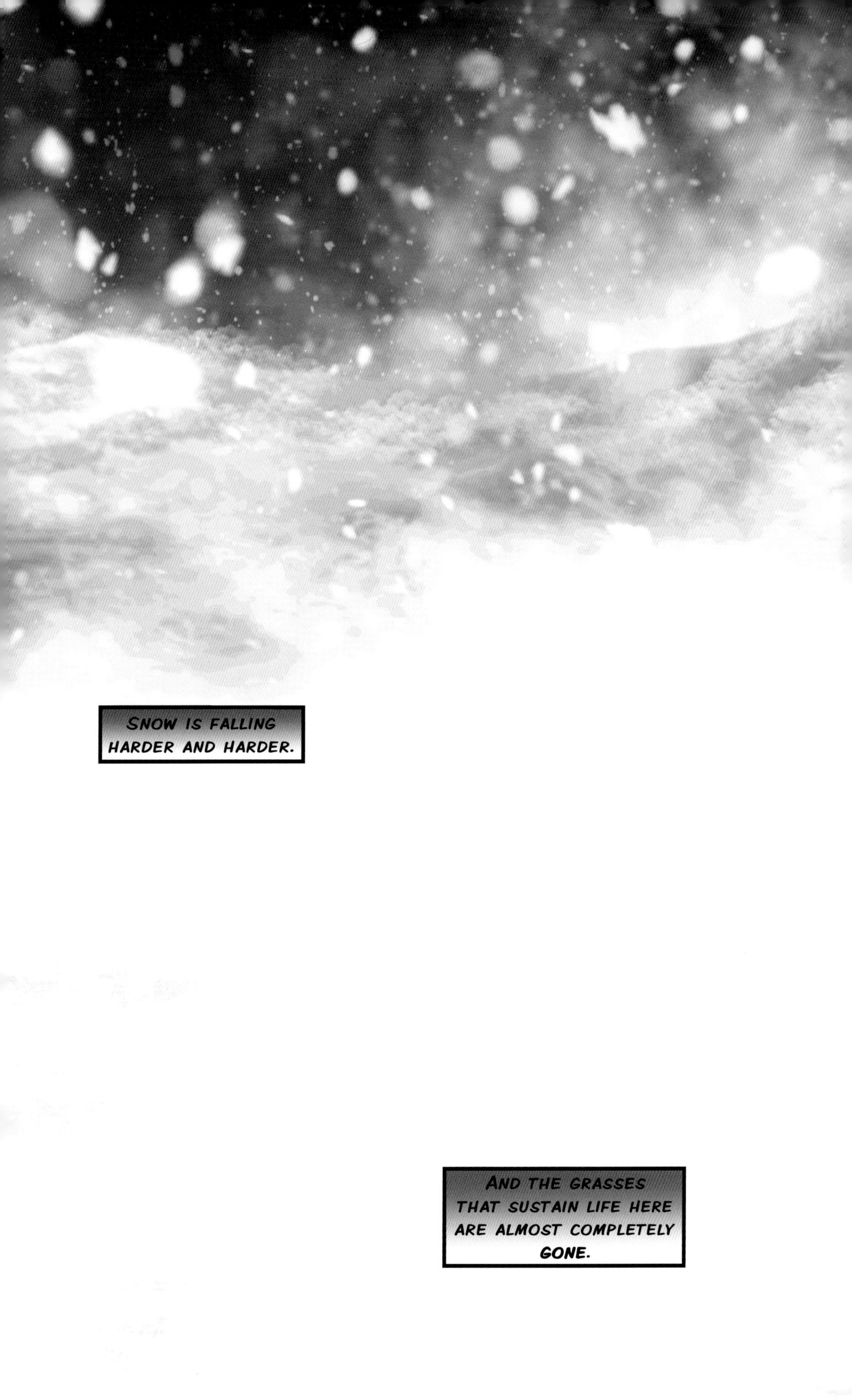
SNOW IS FALLING HARDER AND HARDER.
AND THE GRASSES THAT SUSTAIN LIFE HERE ARE ALMOST COMPLETELY GONE.

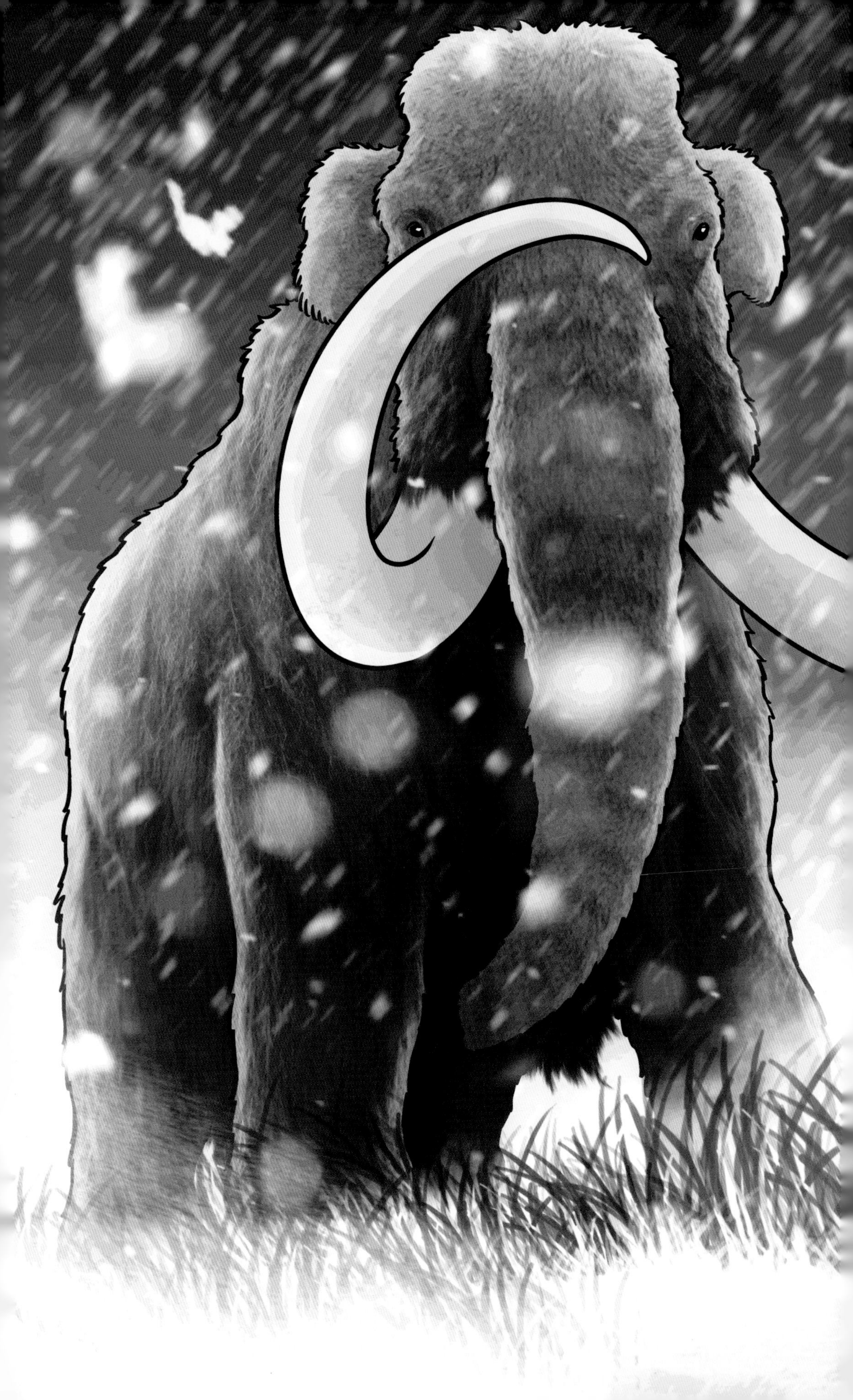

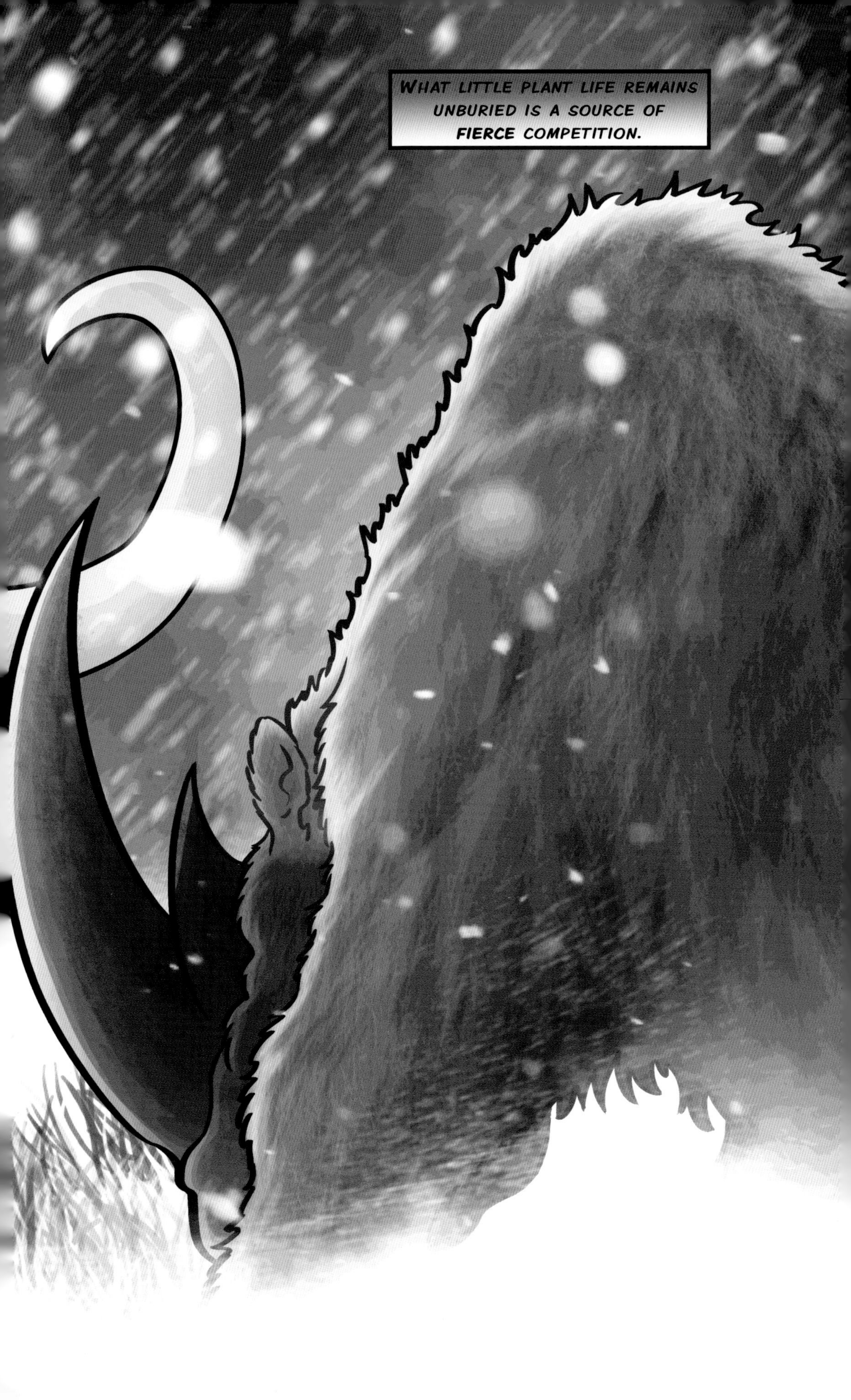
WHAT LITTLE PLANT LIFE REMAINS UNBURIED IS A SOURCE OF FIERCE COMPETITION.

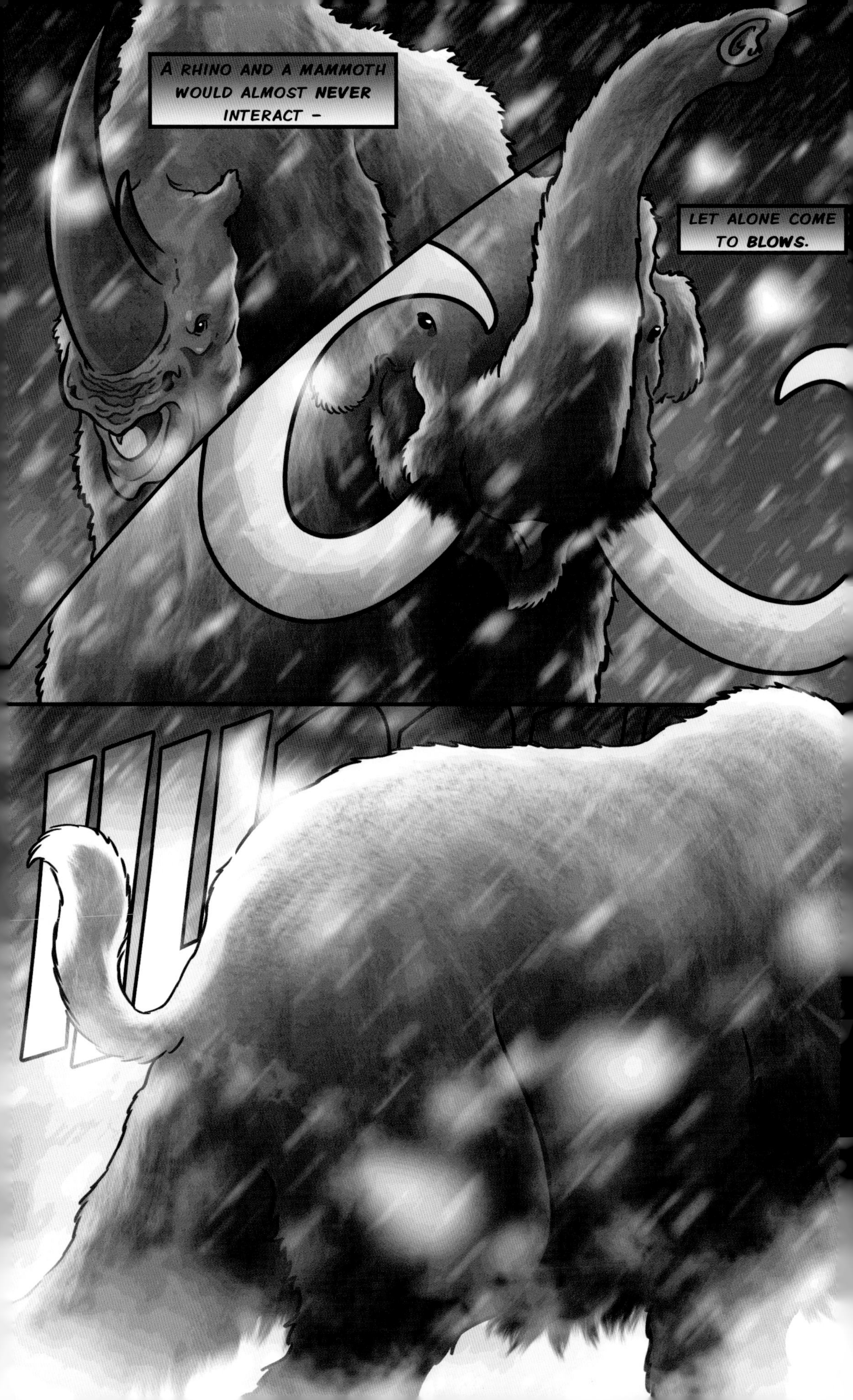
A RHINO AND A MAMMOTH WOULD ALMOST NEVER INTERACT -
LET ALONE COME TO BLOWS.

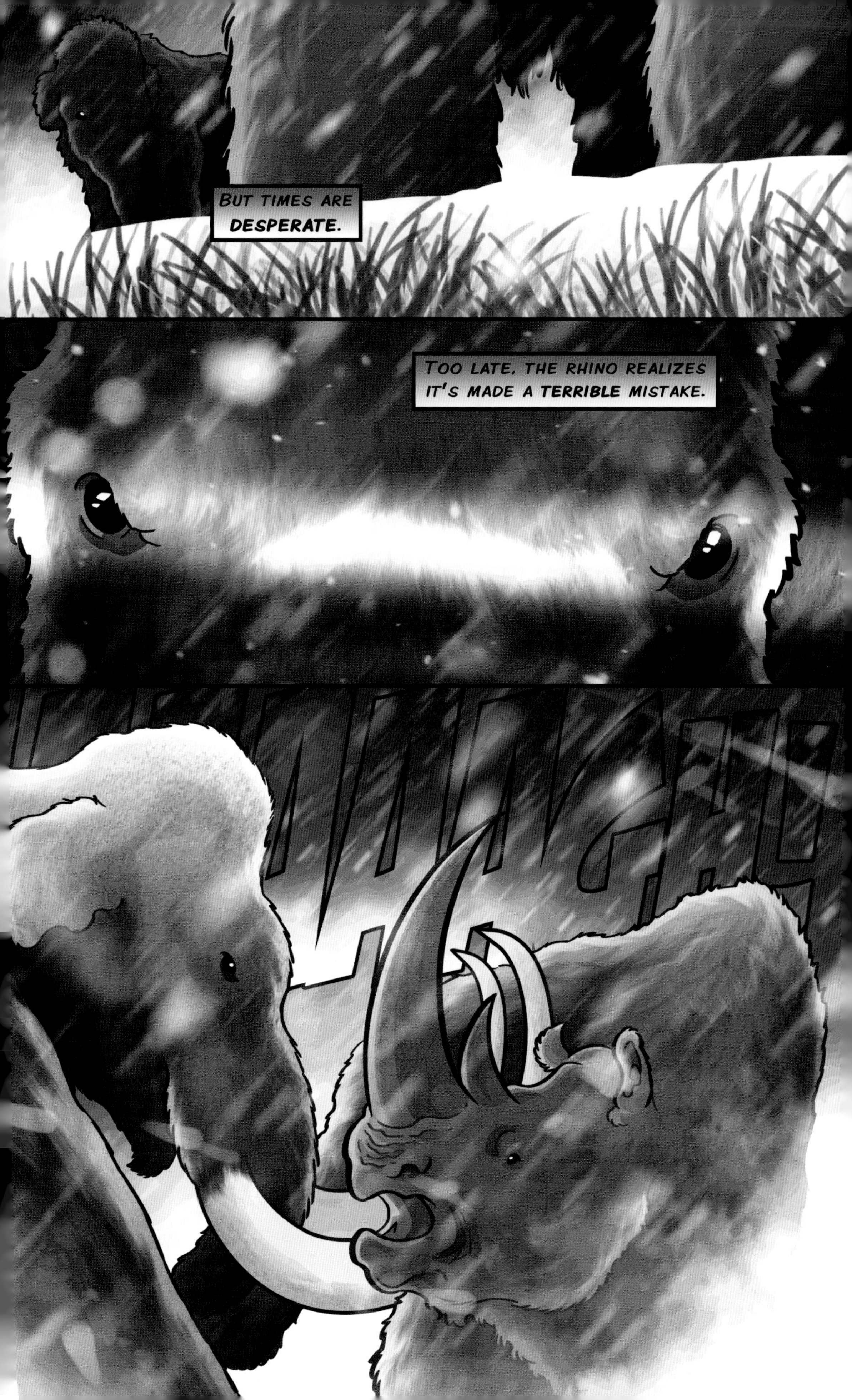
BUT TIMES ARE DESPERATE.
TOO LATE, THE RHINO REALIZES IT'S MADE A TERRIBLE MISTAKE.

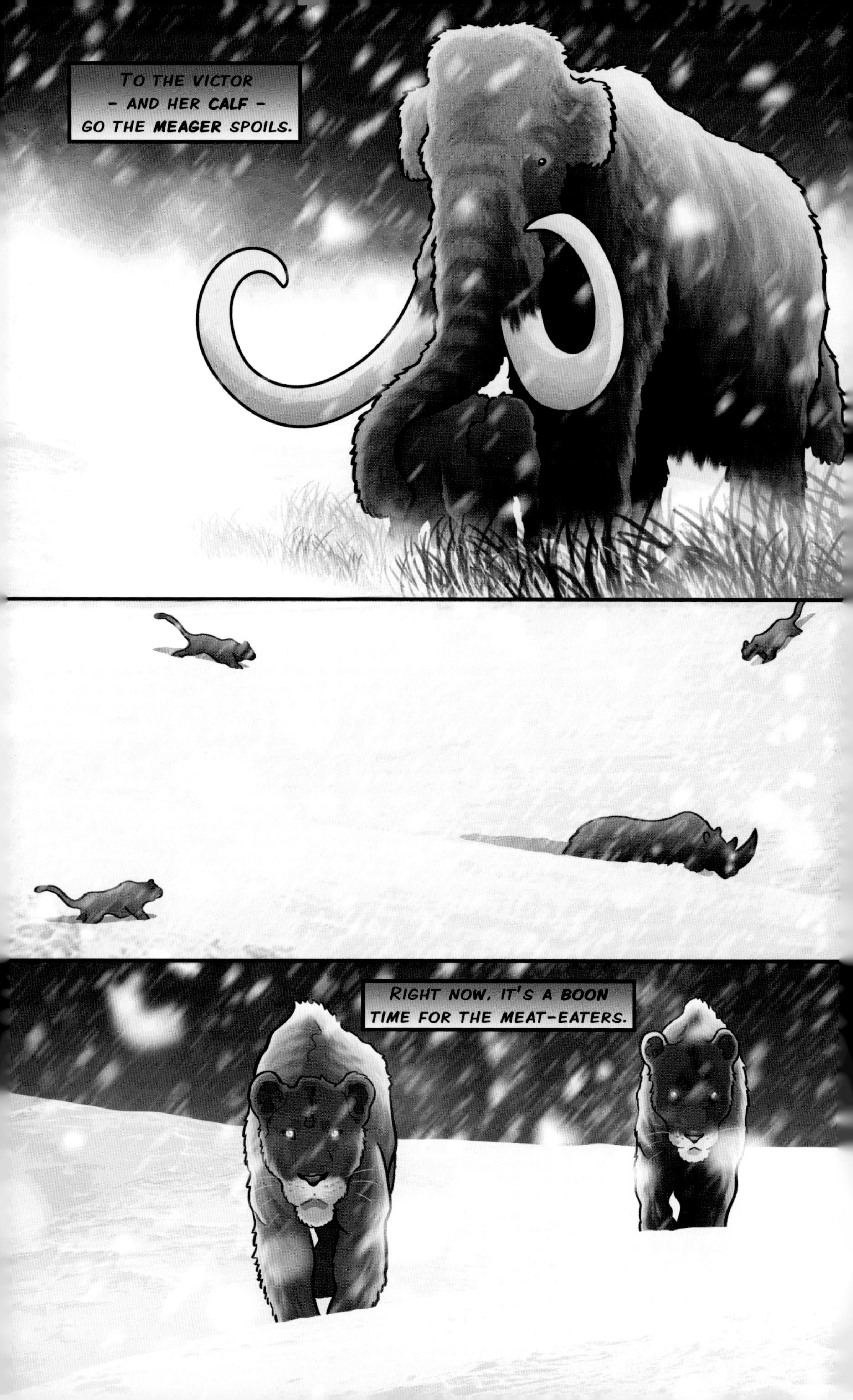
TO THE VICTOR - AND HER CALF - GO THE MEAGER SPOILS.
RIGHT NOW, IT'S A BOON TIME FOR THE MEAT-EATERS.

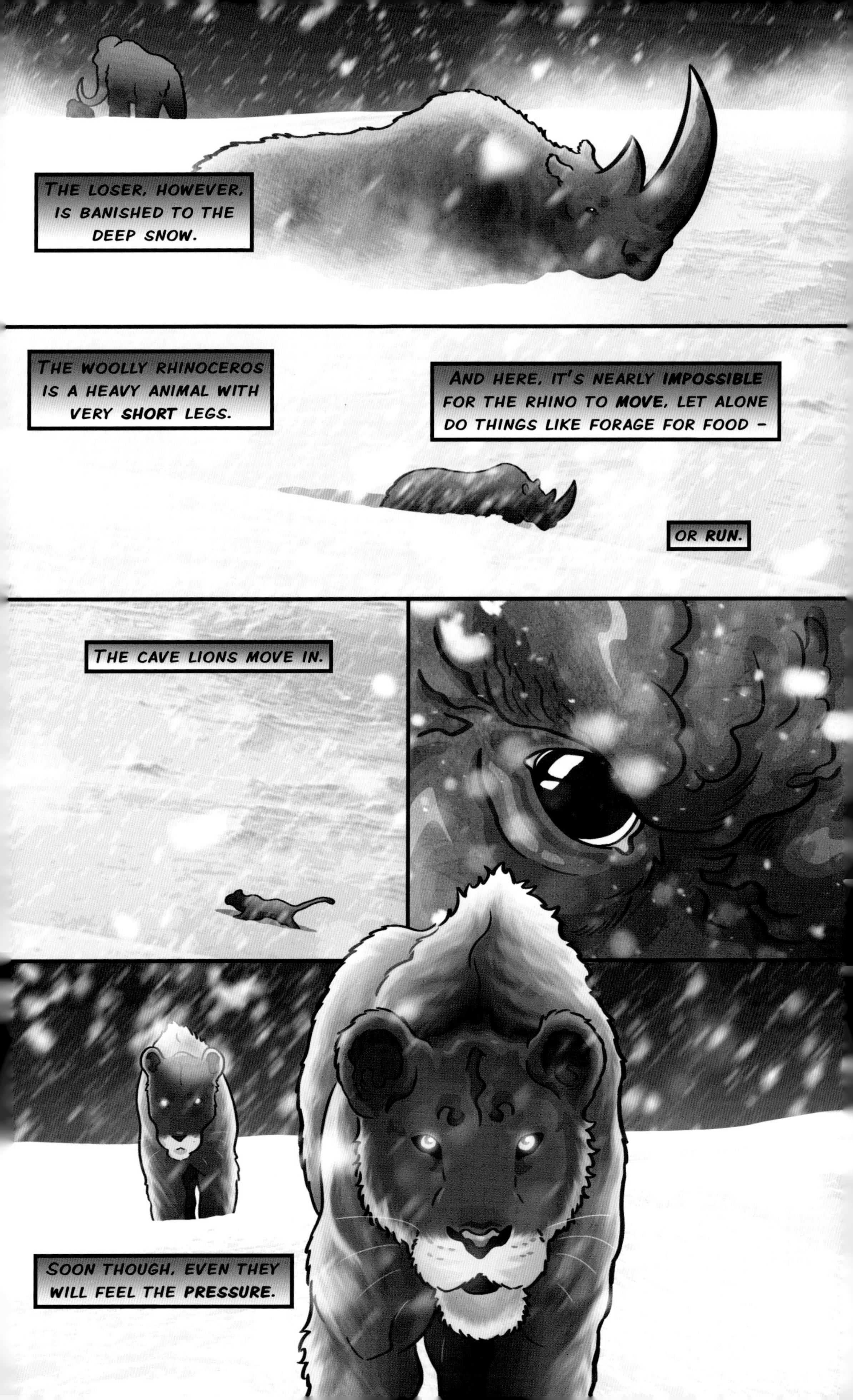
THE LOSER, HOWEVER, IS BANISHED TO THE DEEP SNOW.
THE WOOLLY RHINOCEROS IS A HEAVY ANIMAL WITH VERY SHORT LEGS.
AND HERE, IT'S NEARLY IMPOSSIBLE FOR THE RHINO TO MOVE, LET ALONE DO THINGS LIKE FORAGE FOR FOOD –
OR RUN.
THE CAVE LIONS MOVE IN.
SOON THOUGH, EVEN THEY WILL FEEL THE PRESSURE.

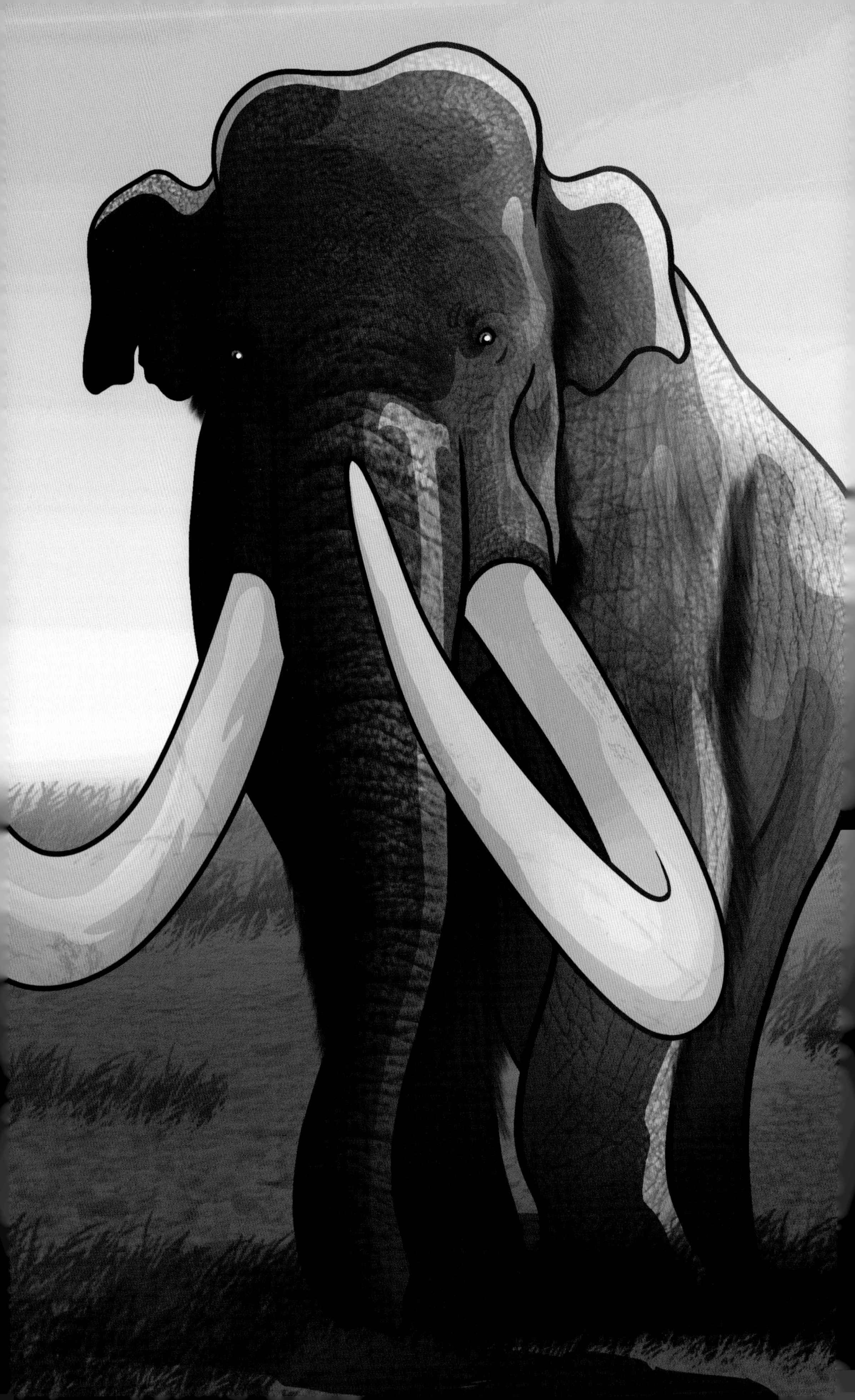

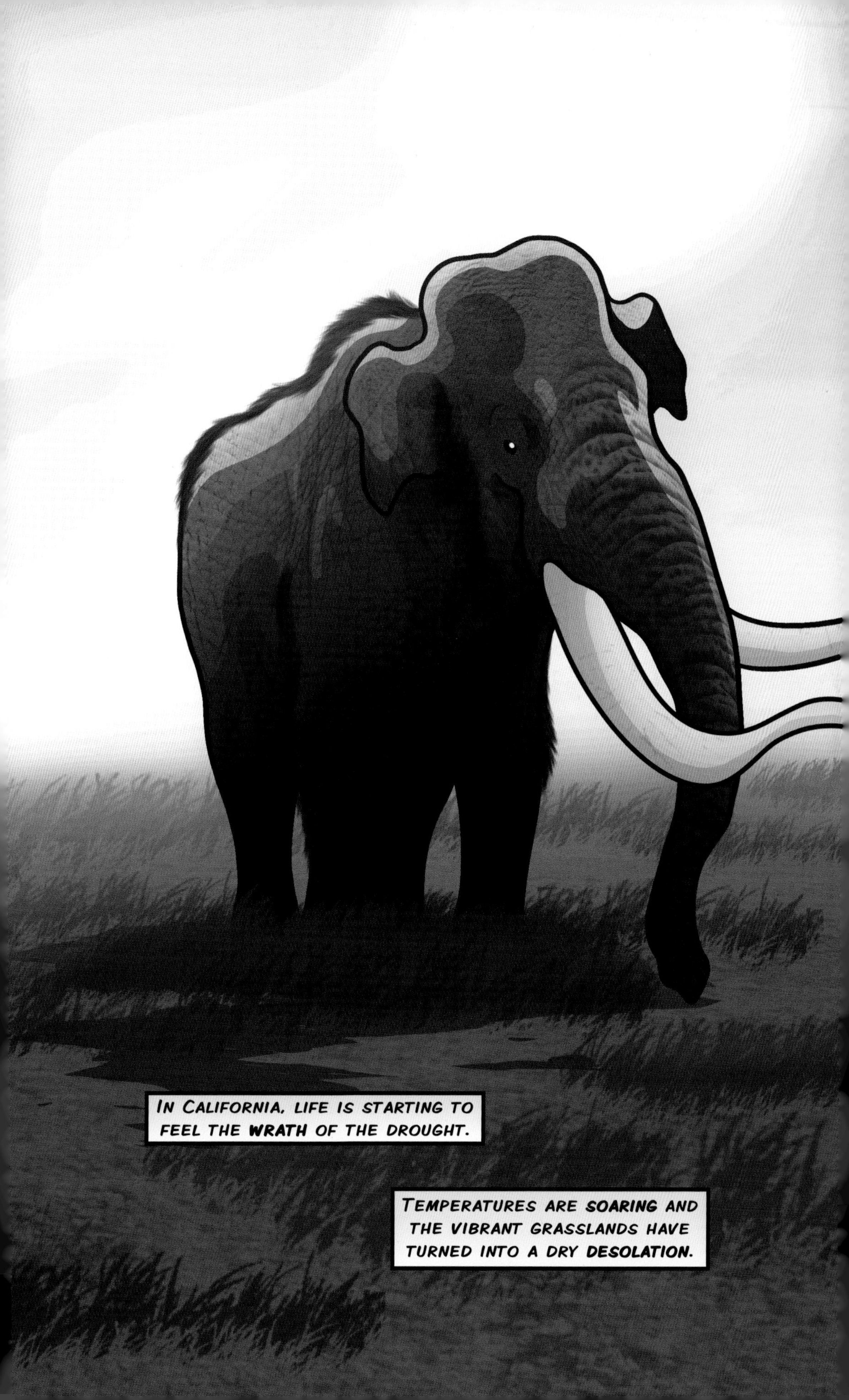
In California, life is starting to feel the wrath of the drought.
Temperatures are soaring and the vibrant grasslands have turned into a dry desolation.

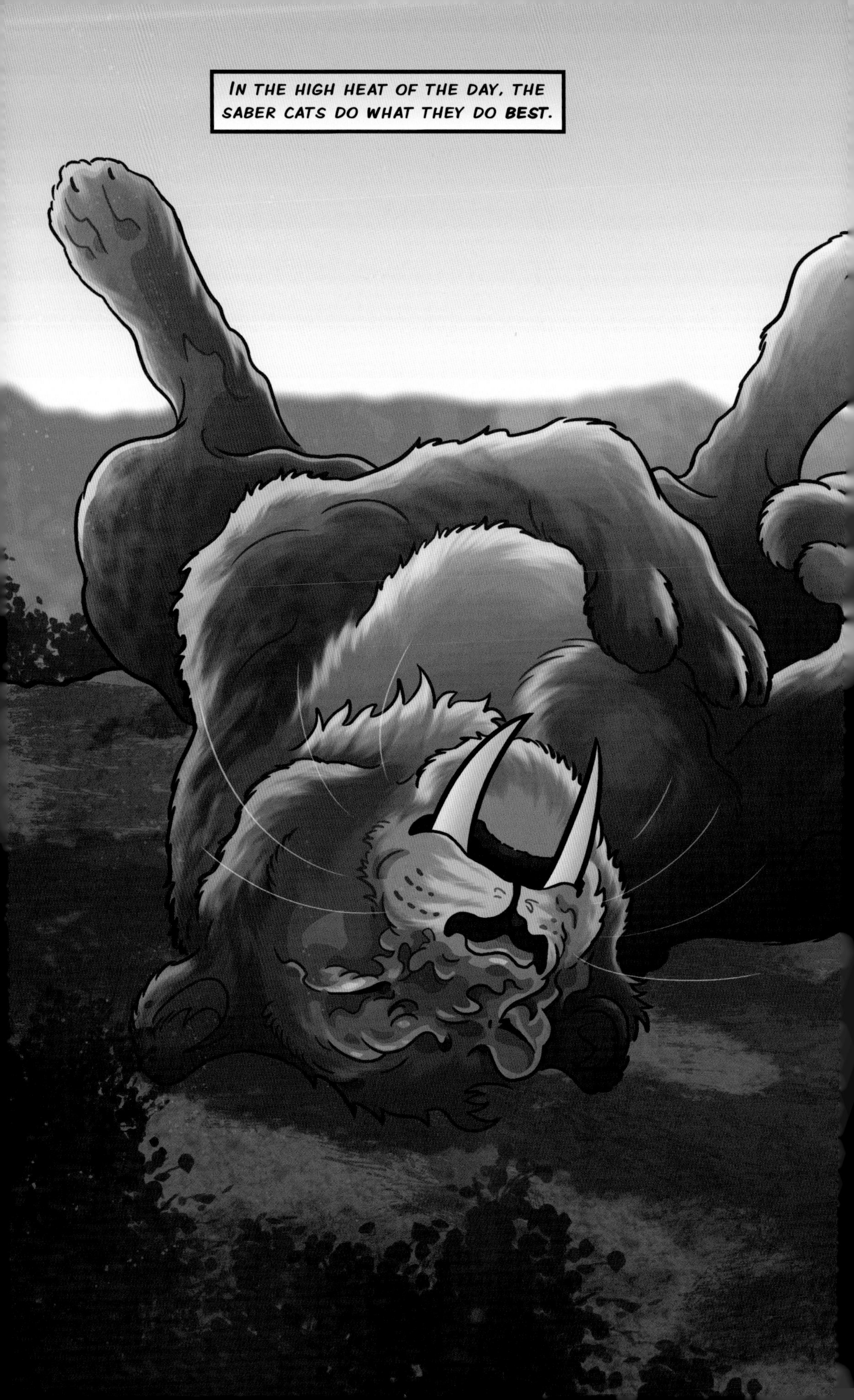
IN THE HIGH HEAT OF THE DAY, THE SABER CATS DO WHAT THEY DO BEST.

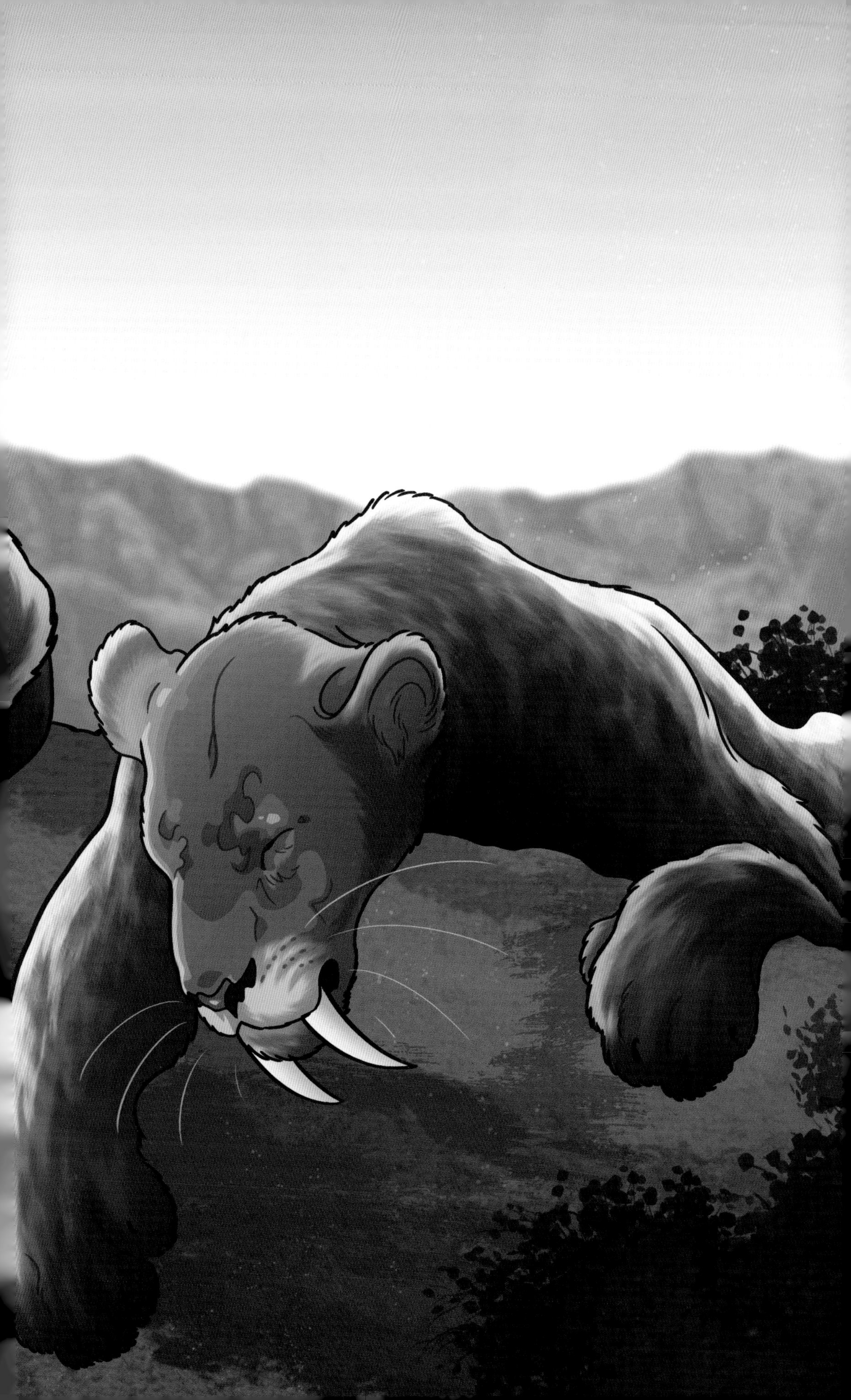

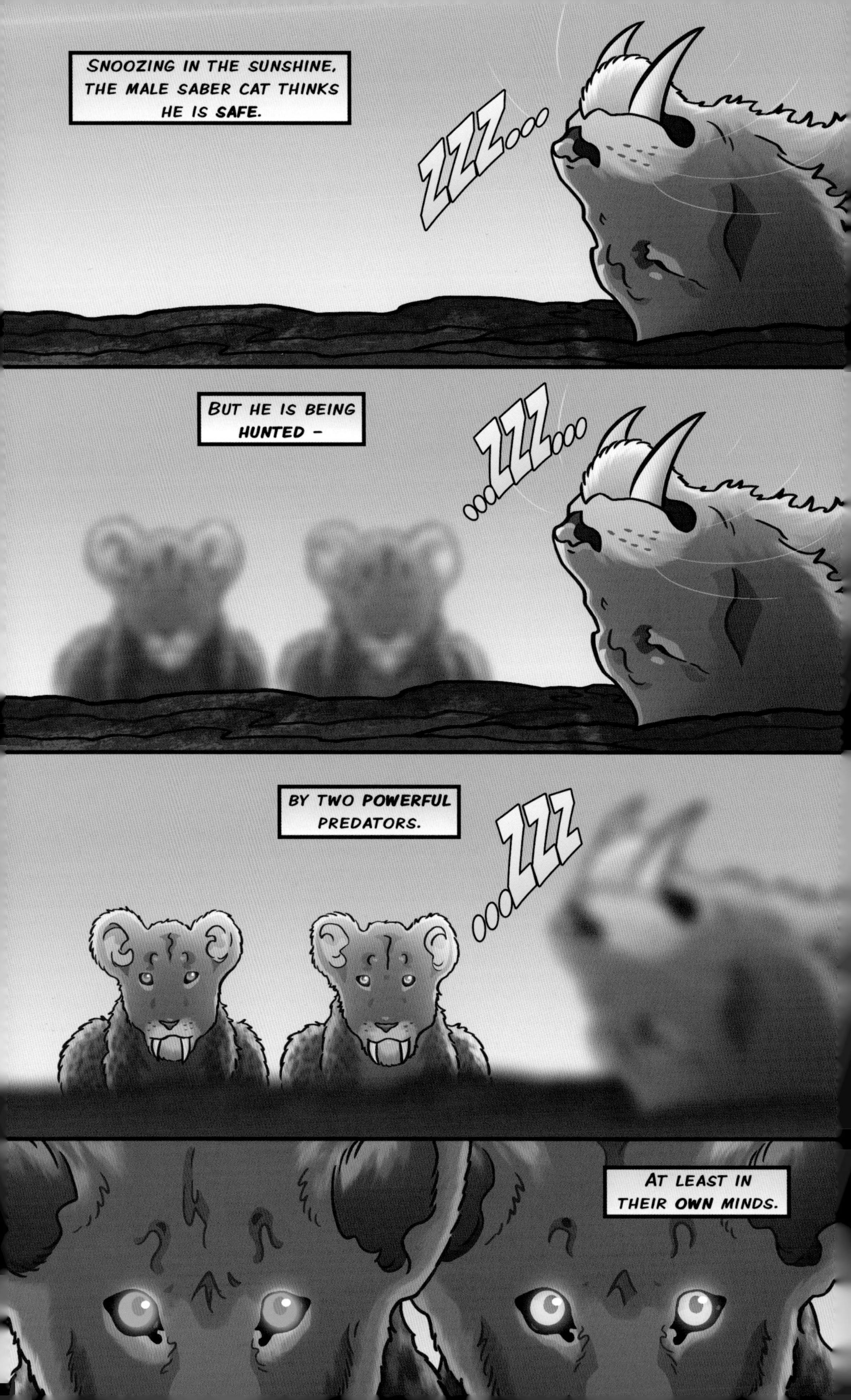
SNOOZING IN THE SUNSHINE, THE MALE SABER CAT THINKS HE IS SAFE.
ZZZ...
BUT HE IS BEING HUNTED –
...ZZZ...
BY TWO POWERFUL PREDATORS.
...ZZZ
AT LEAST IN THEIR OWN MINDS.

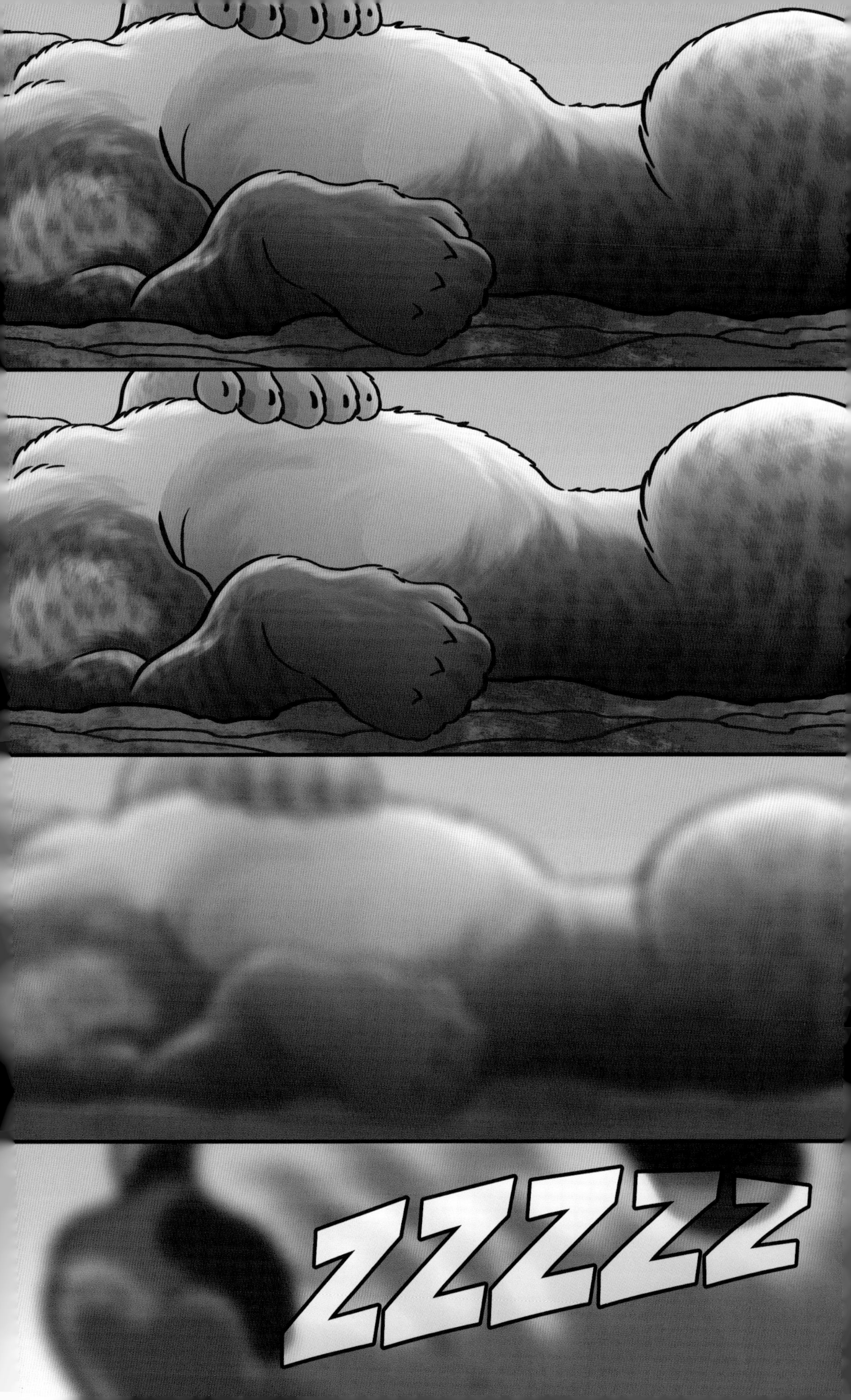
ZZZZZ

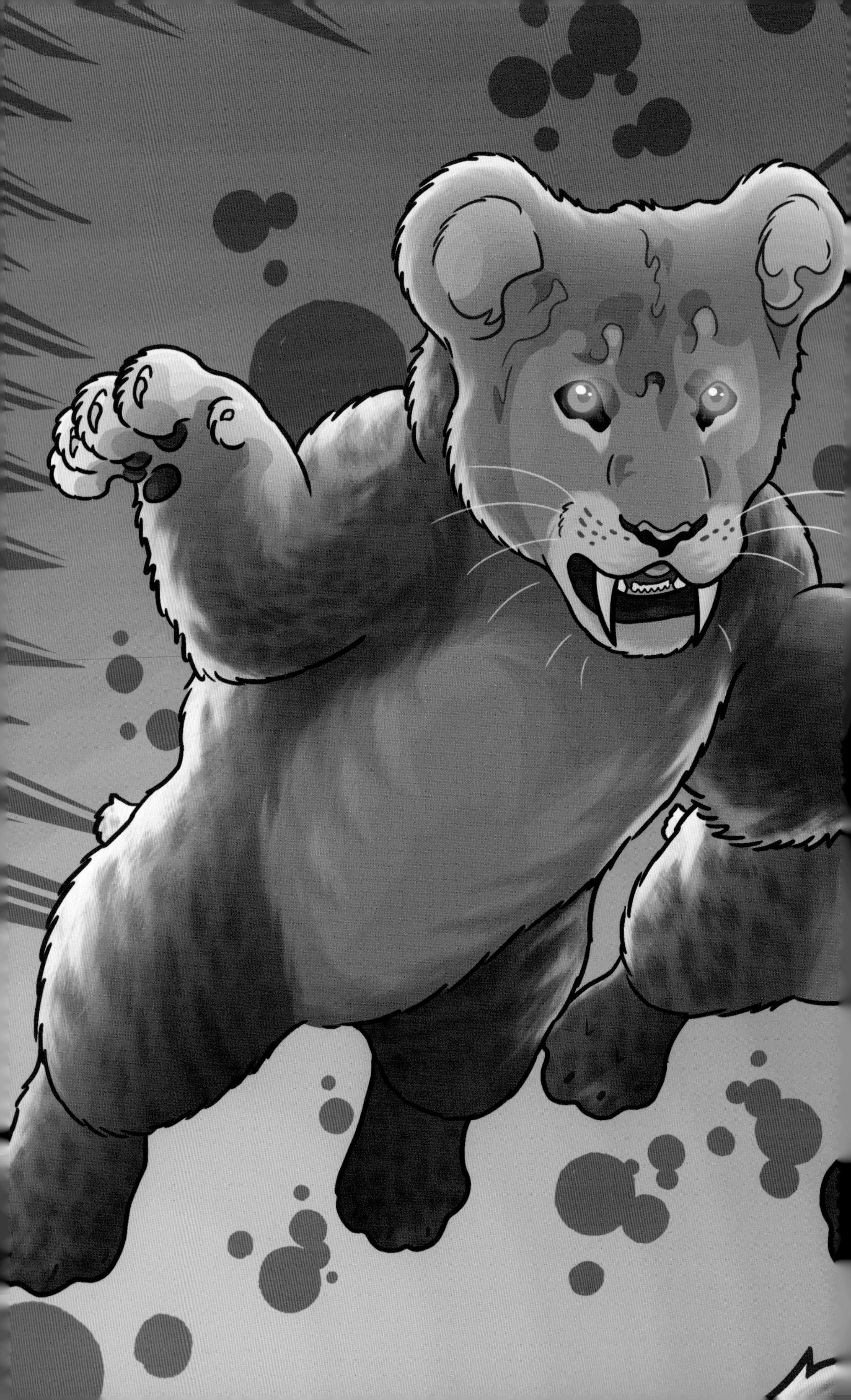

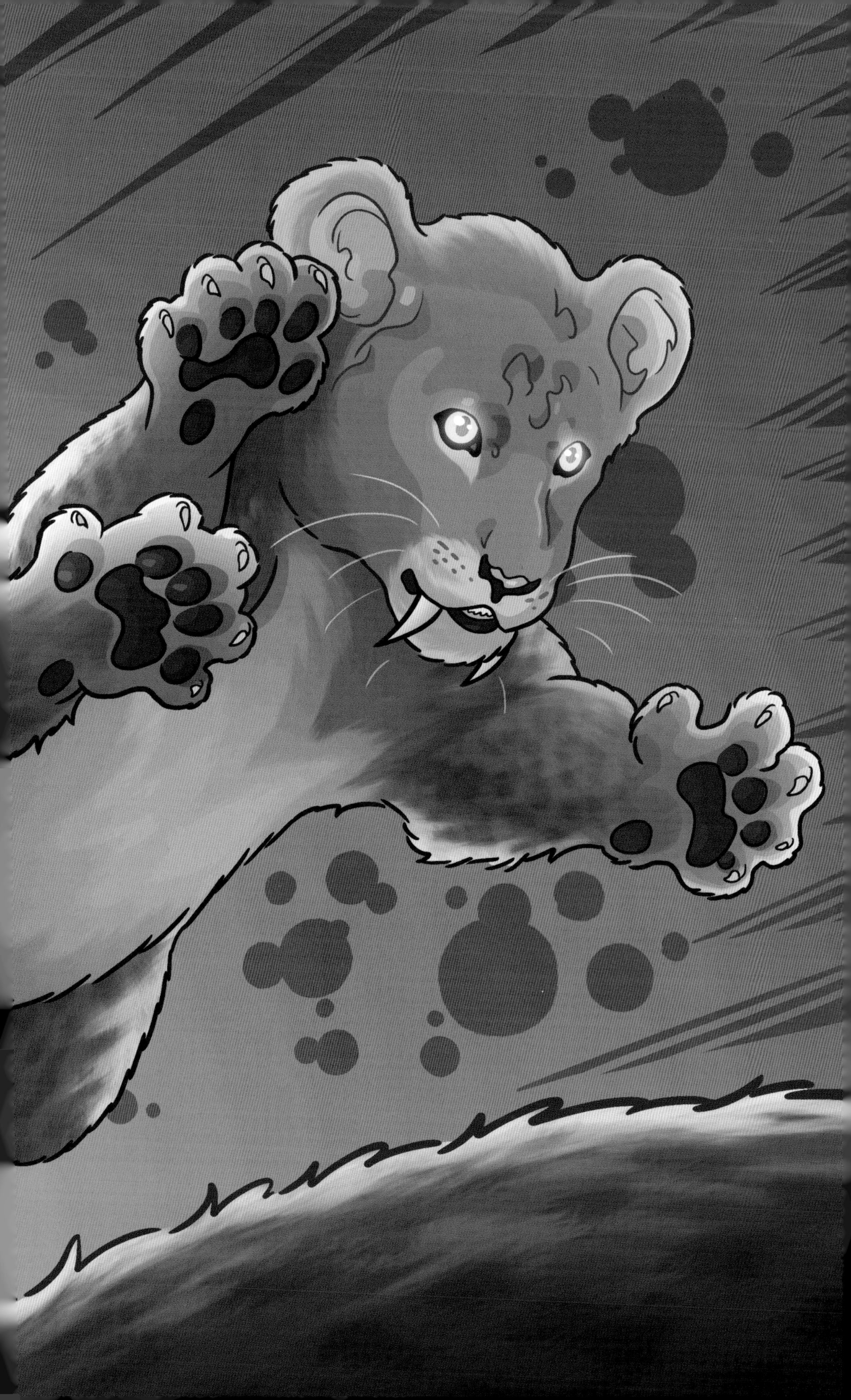

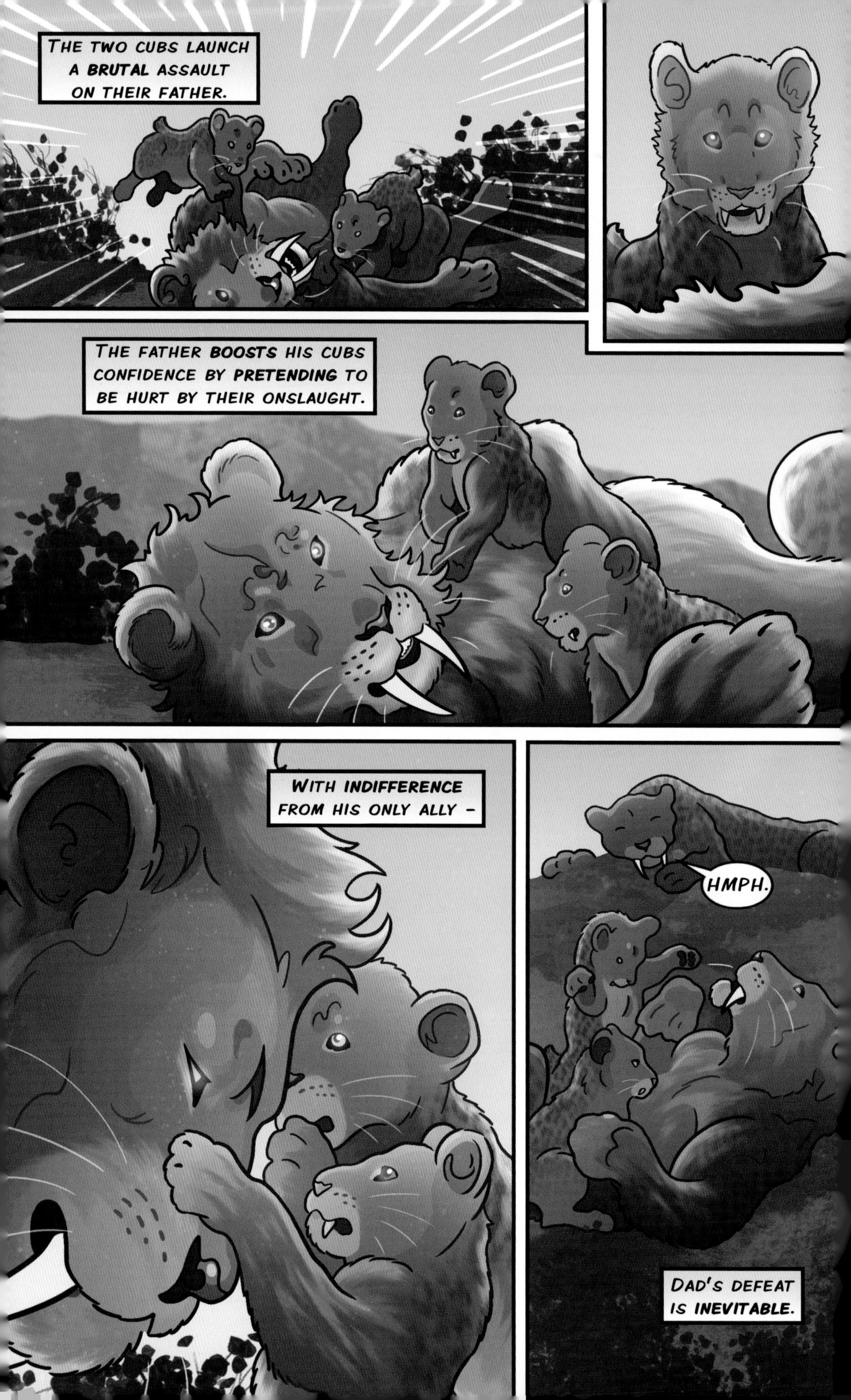
THE TWO CUBS LAUNCH A BRUTAL ASSAULT ON THEIR FATHER.
THE FATHER BOOSTS HIS CUBS CONFIDENCE BY PRETENDING TO BE HURT BY THEIR ONSLAUGHT.
WITH INDIFFERENCE FROM HIS ONLY ALLY -
HMPH.
DAD'S DEFEAT IS INEVITABLE.

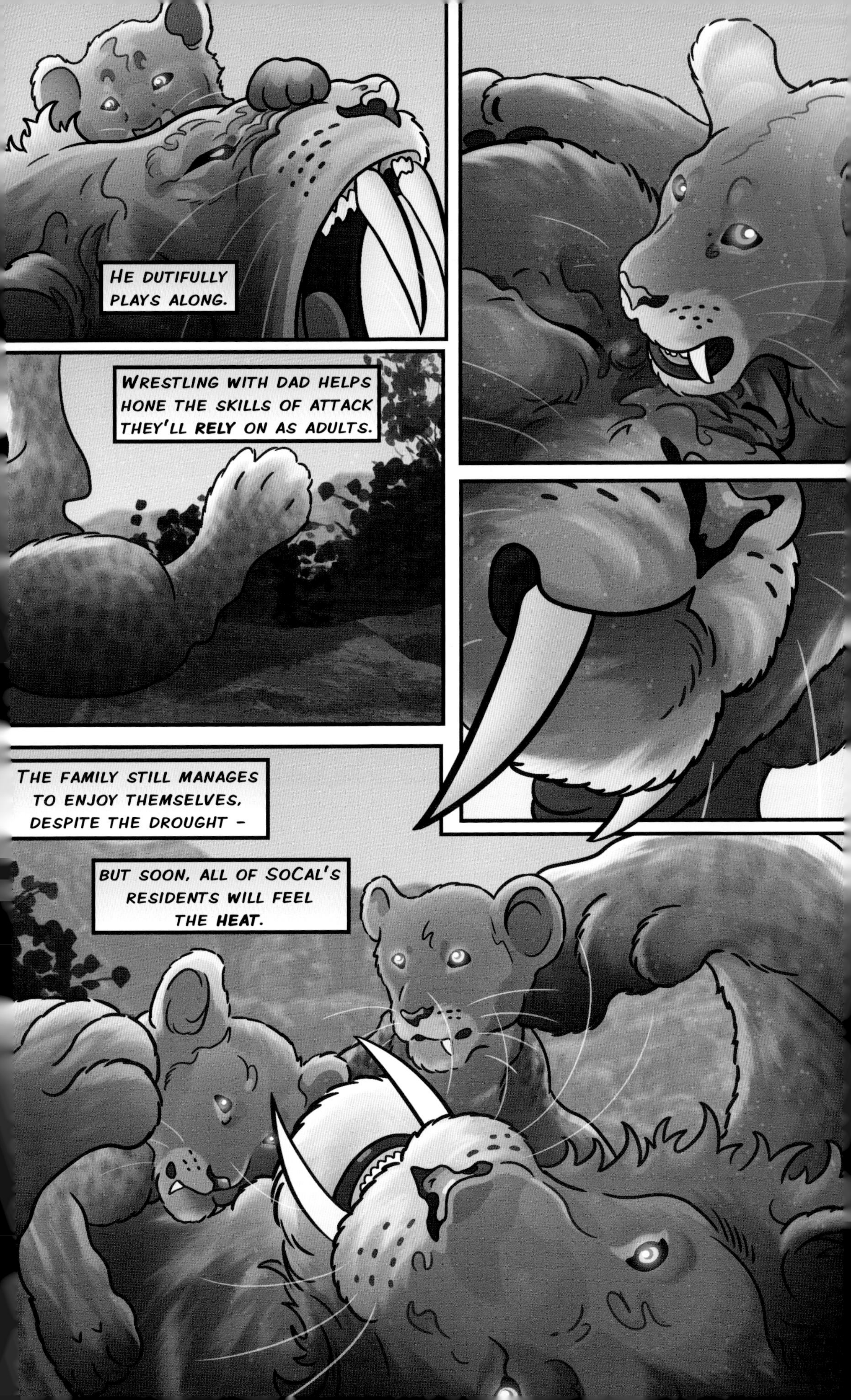
HE DUTIFULLY PLAYS ALONG.
WRESTLING WITH DAD HELPS HONE THE SKILLS OF ATTACK THEY'LL RELY ON AS ADULTS.
THE FAMILY STILL MANAGES TO ENJOY THEMSELVES, DESPITE THE DROUGHT –
BUT SOON, ALL OF SOCAL'S RESIDENTS WILL FEEL THE HEAT.

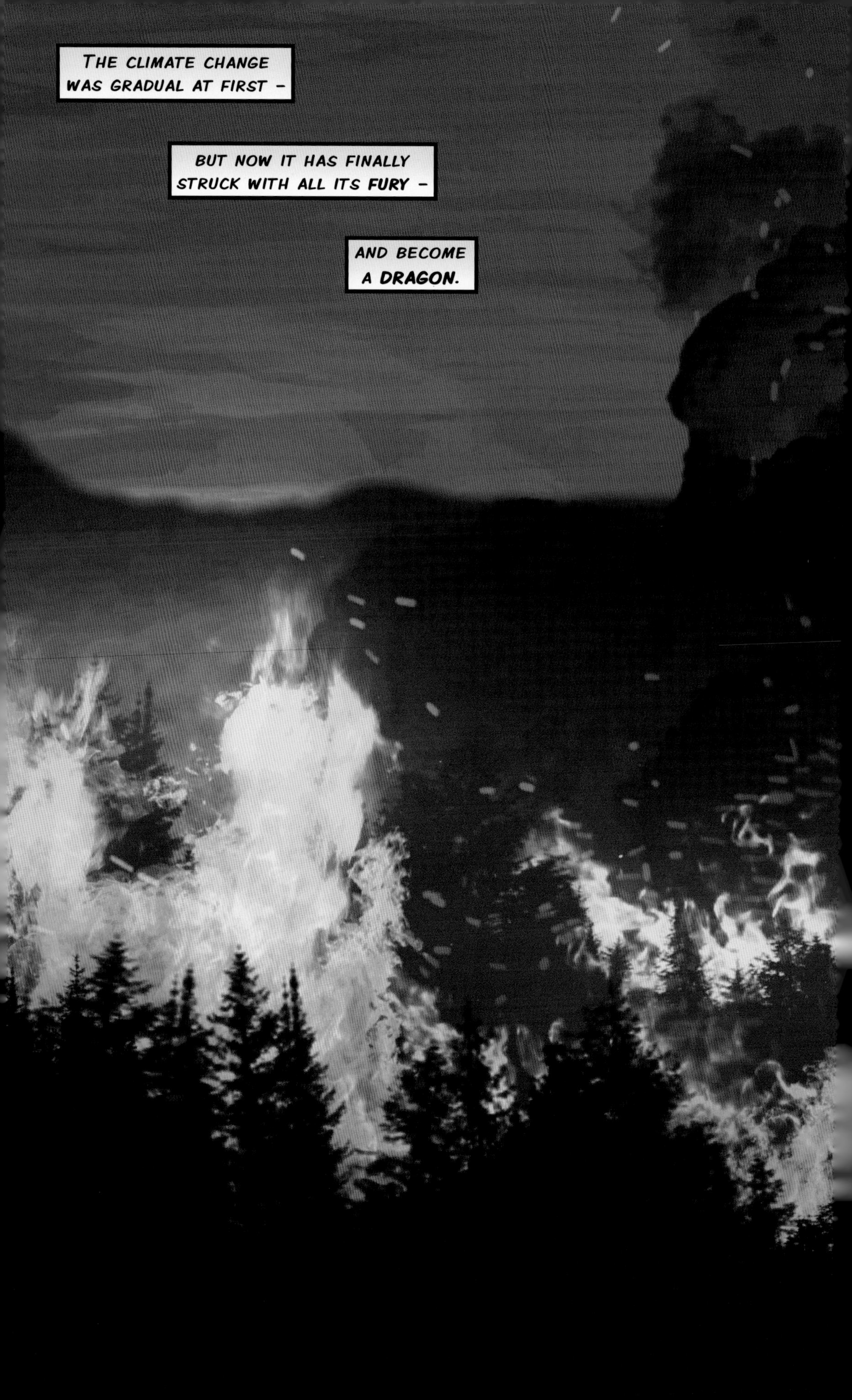
THE CLIMATE CHANGE WAS GRADUAL AT FIRST –
BUT NOW IT HAS FINALLY STRUCK WITH ALL ITS FURY –
AND BECOME A DRAGON.

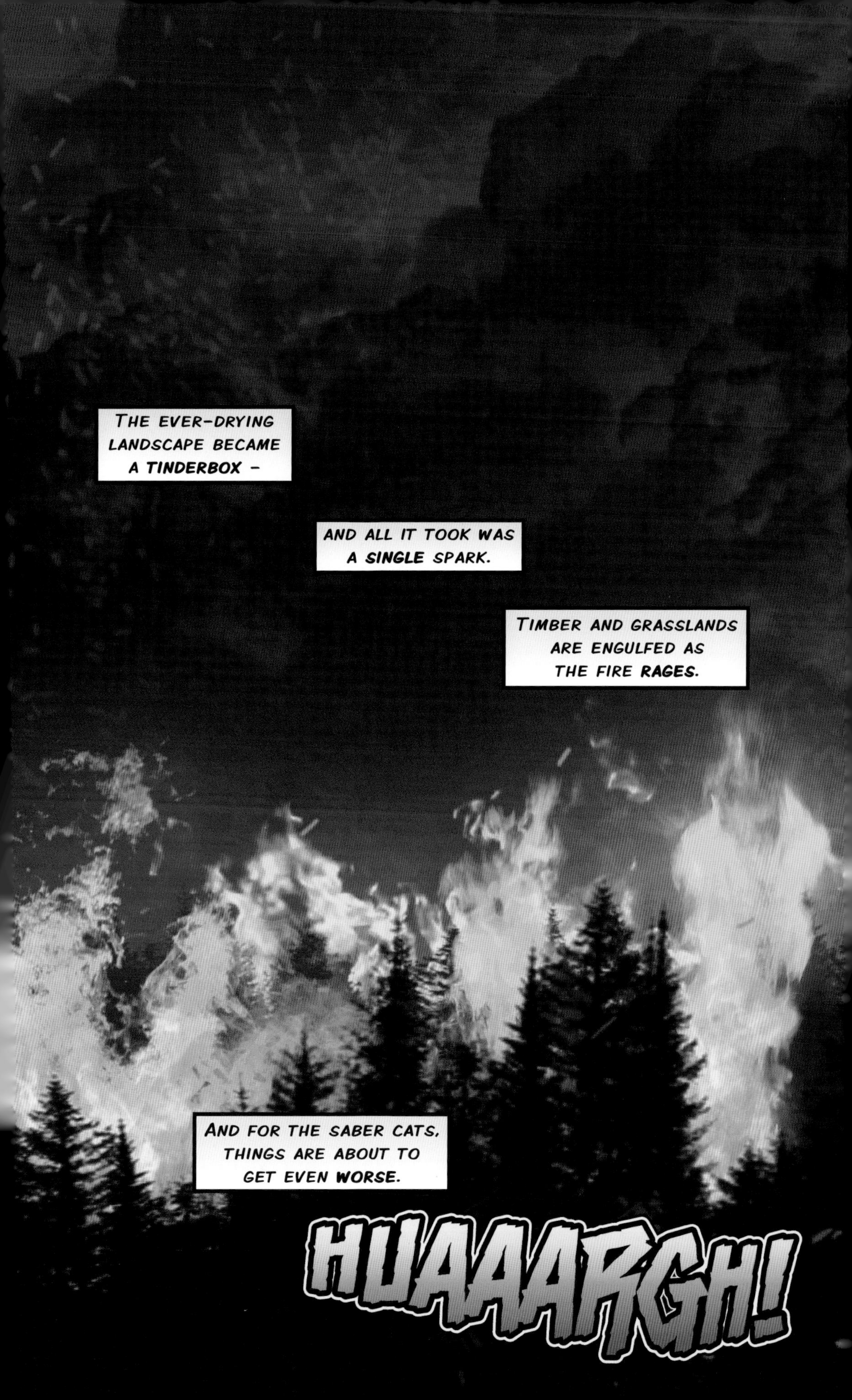
THE EVER-DRYING LANDSCAPE BECAME A TINDERBOX –
AND ALL IT TOOK WAS A SINGLE SPARK.
TIMBER AND GRASSLANDS ARE ENGULFED AS THE FIRE RAGES.
AND FOR THE SABER CATS, THINGS ARE ABOUT TO GET EVEN WORSE.
HUAAARGH!

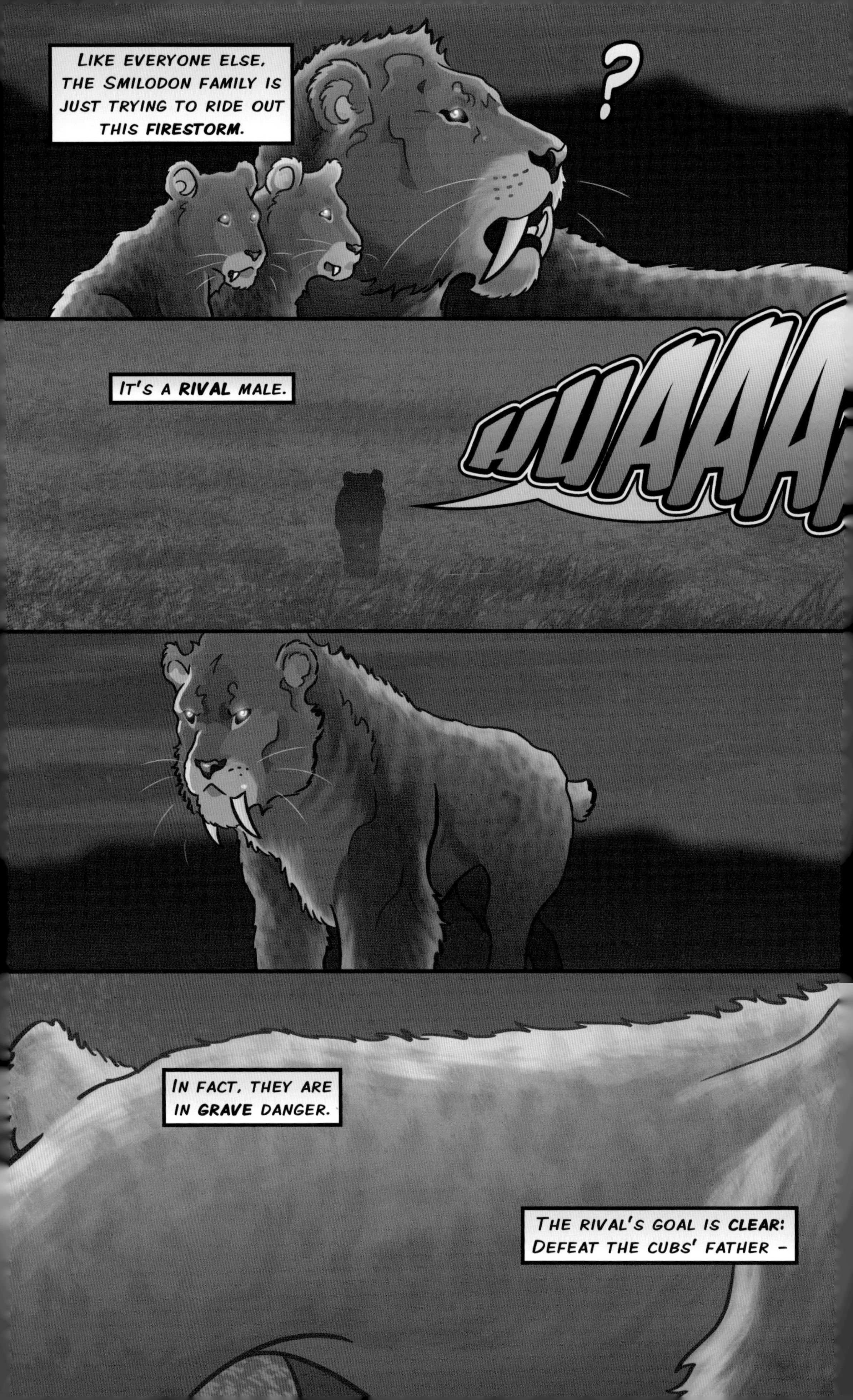
LIKE EVERYONE ELSE, THE SMILODON FAMILY IS JUST TRYING TO RIDE OUT THIS FIRESTORM.
?
IT'S A RIVAL MALE.
WUAAAA
IN FACT, THEY ARE IN GRAVE DANGER.
THE RIVAL'S GOAL IS CLEAR: DEFEAT THE CUBS' FATHER –

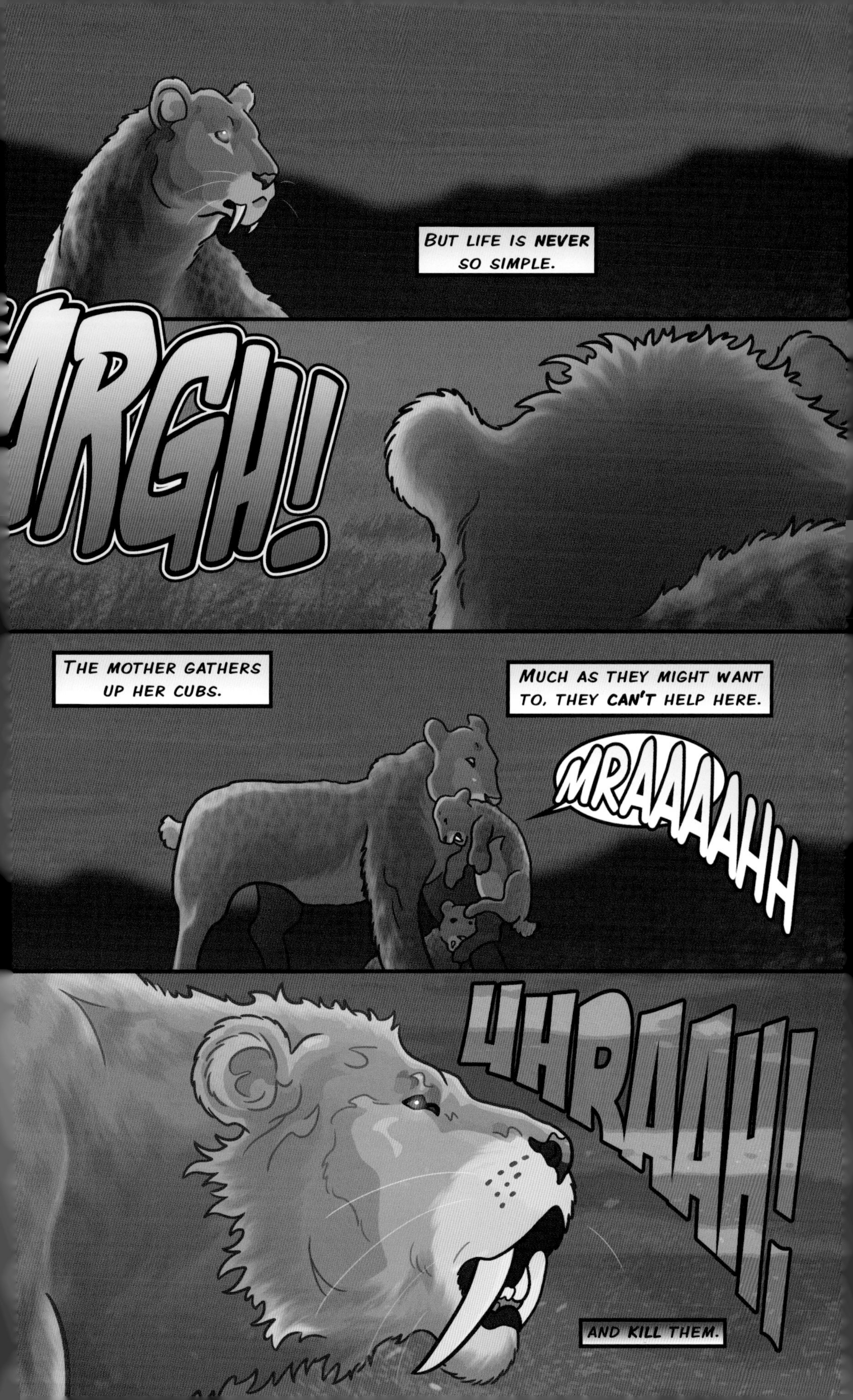
BUT LIFE IS **NEVER** SO SIMPLE.
RGH!
THE MOTHER GATHERS UP HER CUBS.
MUCH AS THEY MIGHT WANT TO, THEY **CAN'T** HELP HERE.
MRAAAAAHH
UHRAAH!
AND KILL THEM.

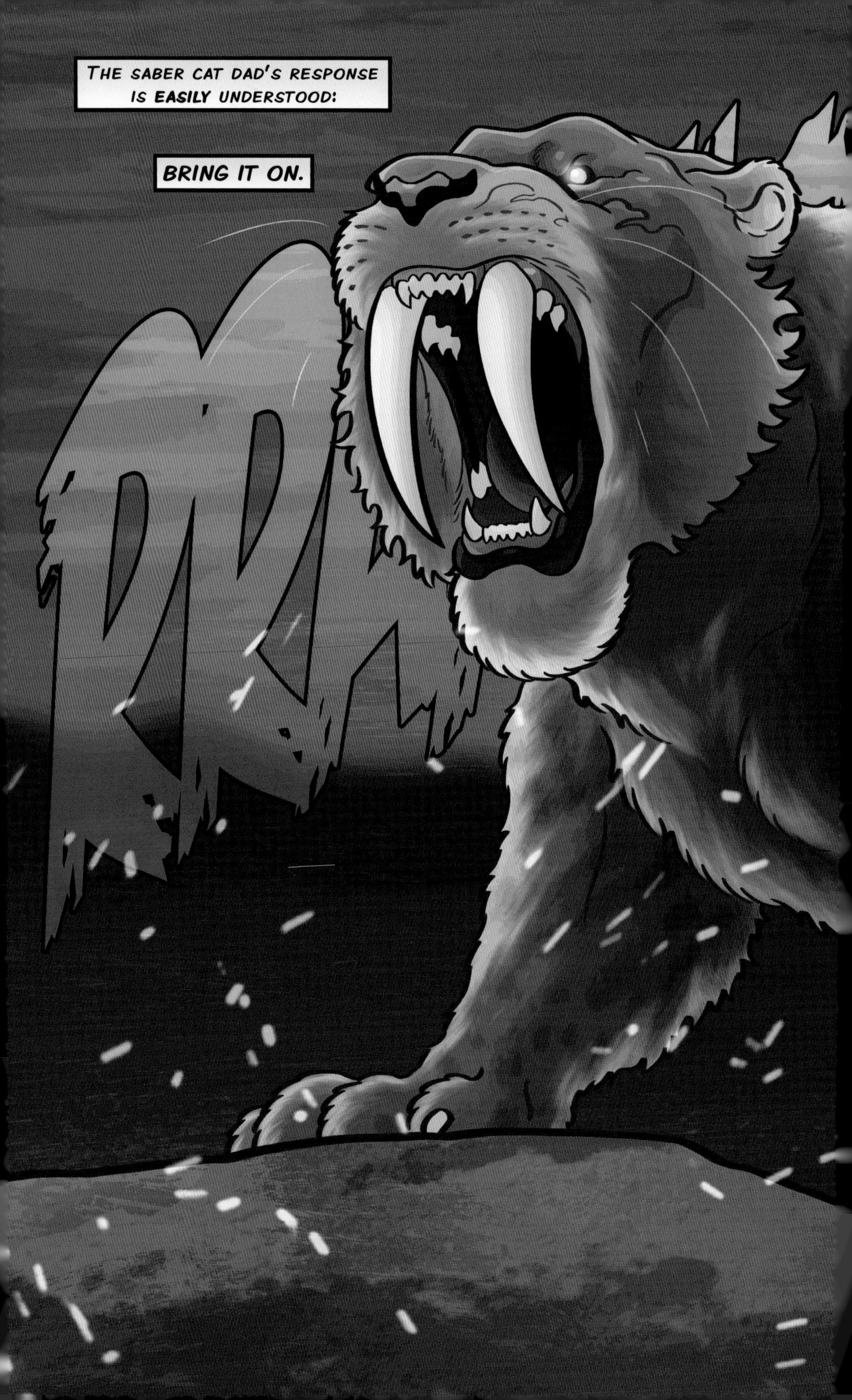
THE SABER CAT DAD'S RESPONSE IS EASILY UNDERSTOOD:
BRING IT ON.

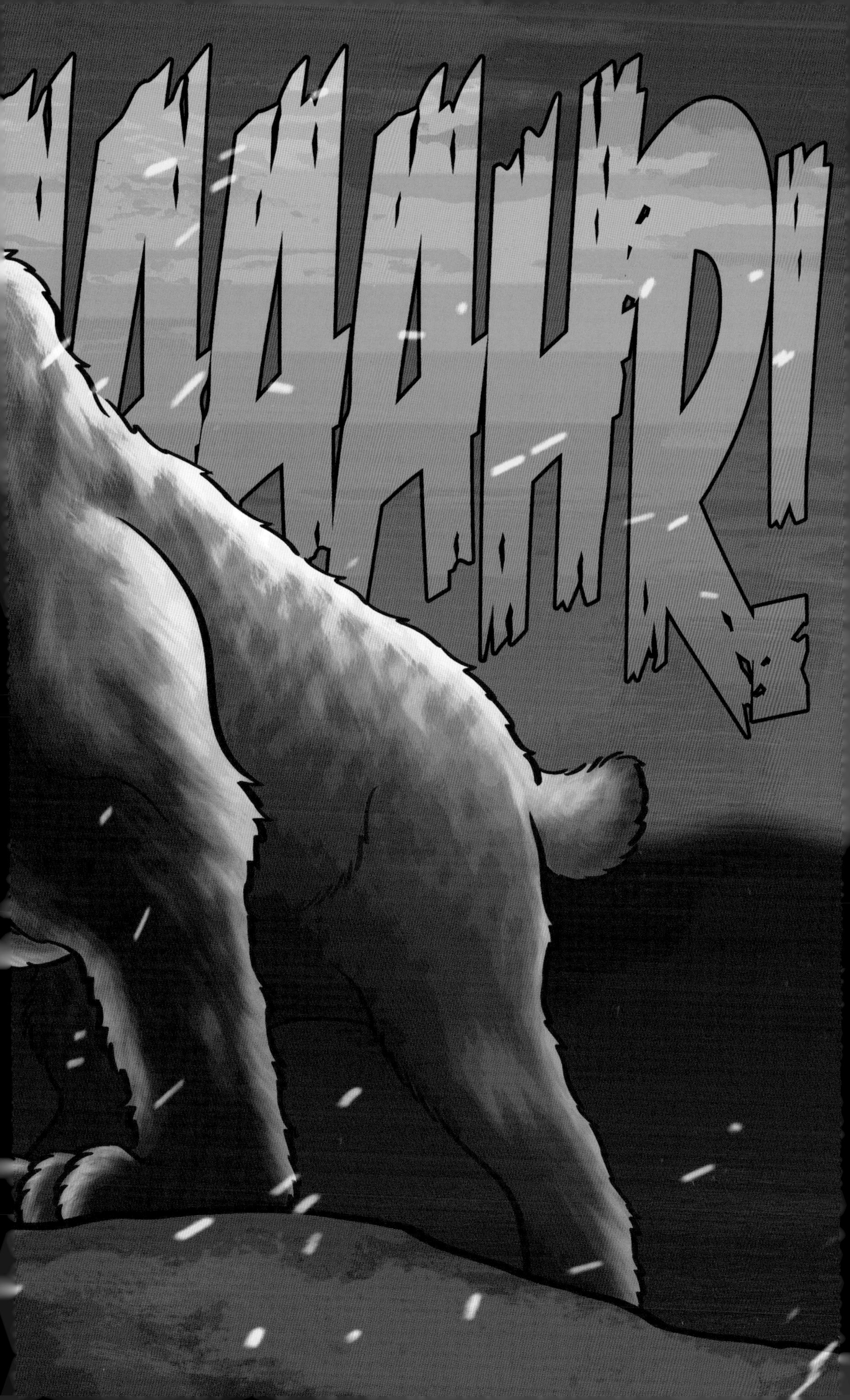
AAAHR!

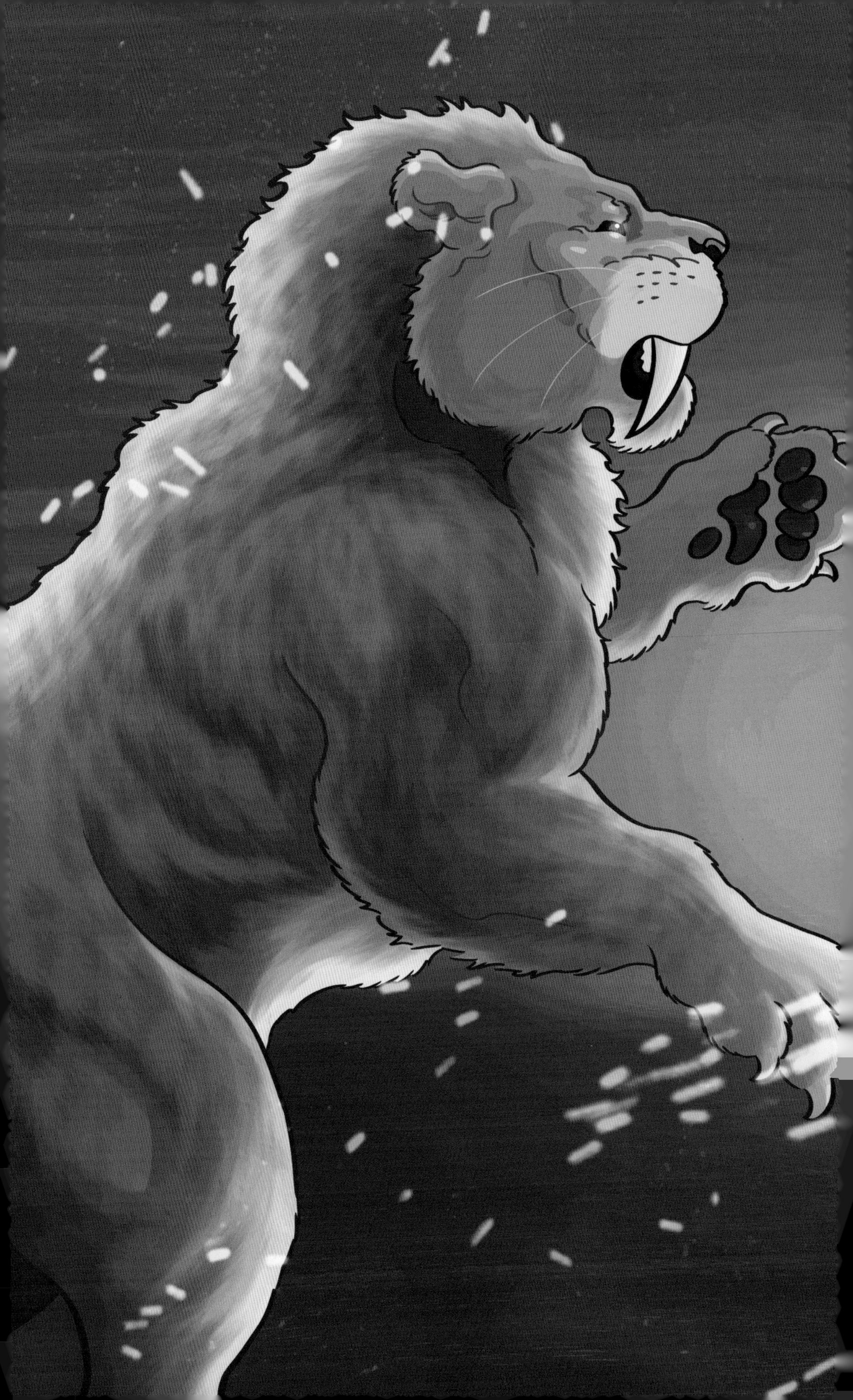

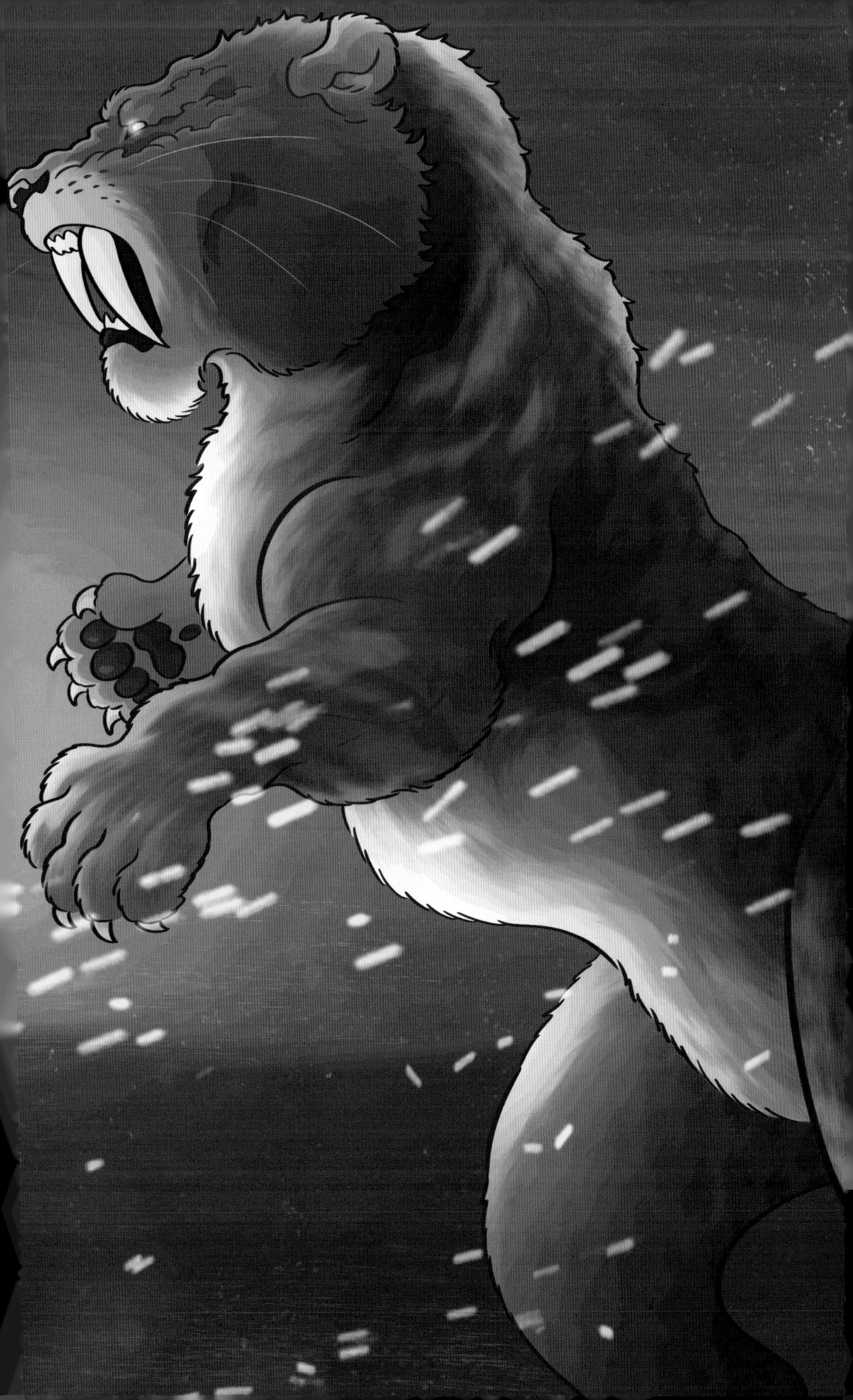

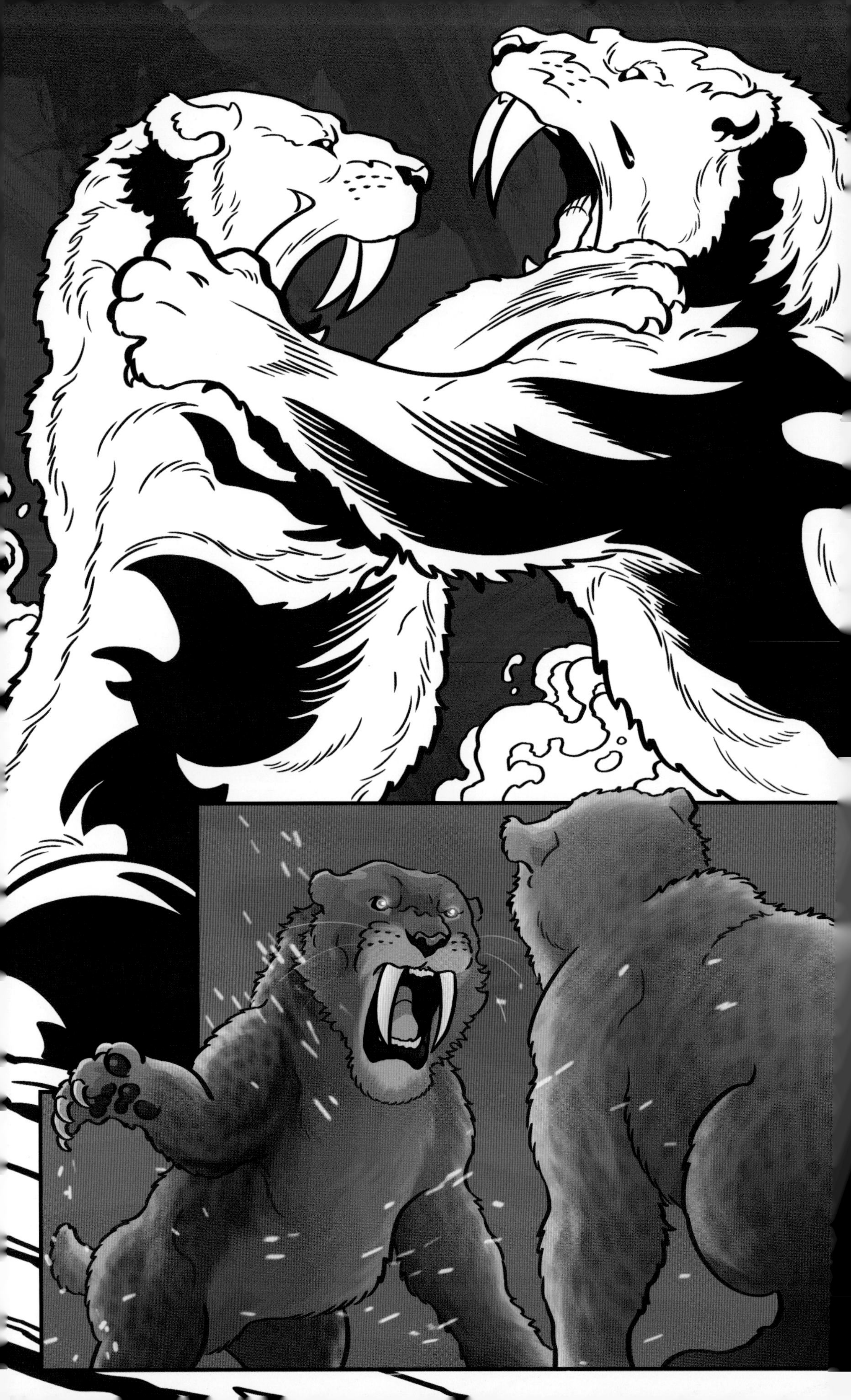

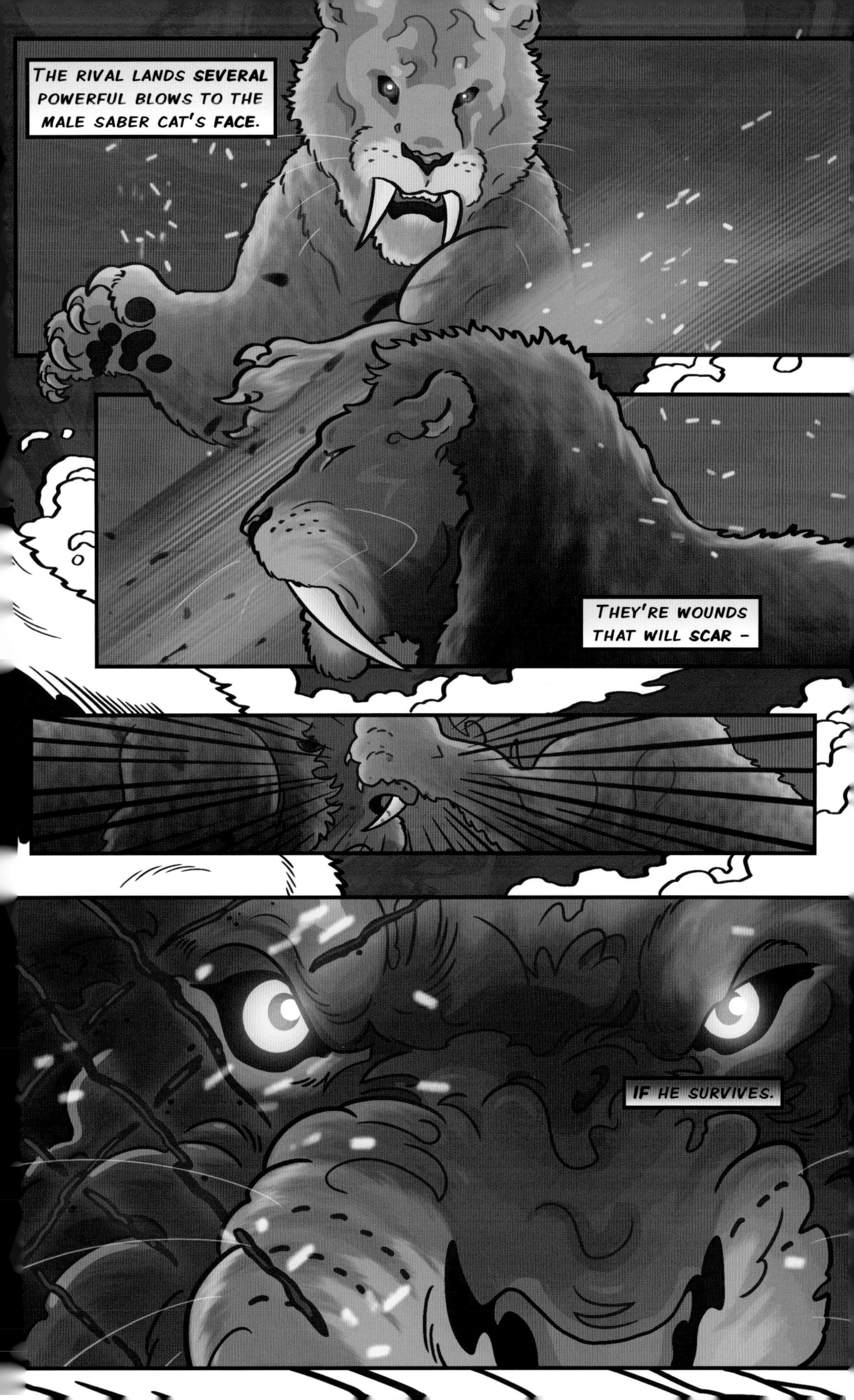
THE RIVAL LANDS SEVERAL POWERFUL BLOWS TO THE MALE SABER CAT'S FACE.
THEY'RE WOUNDS THAT WILL SCAR -
IF HE SURVIVES.

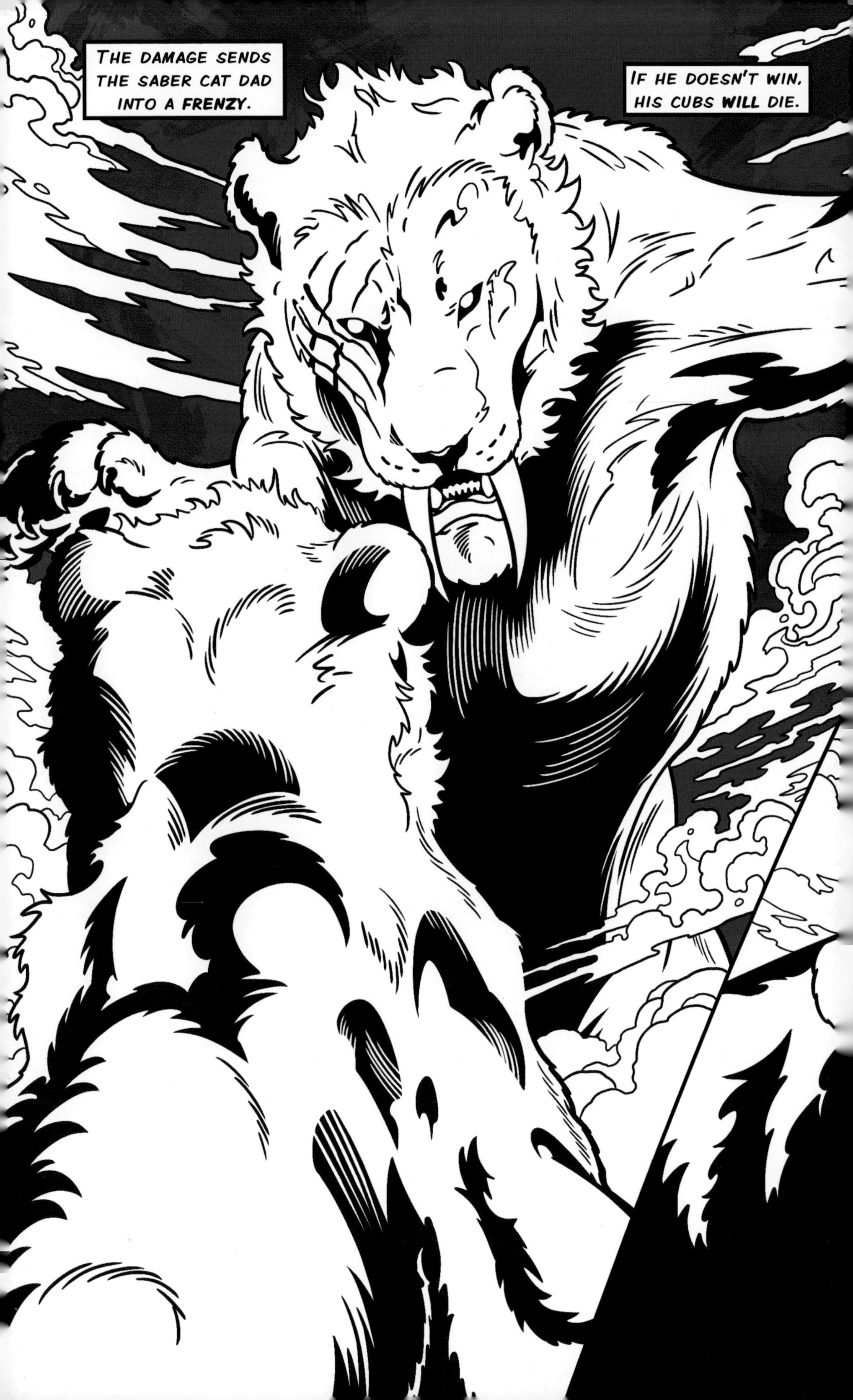
THE DAMAGE SENDS THE SABER CAT DAD INTO A **FRENZY**.
IF HE DOESN'T WIN, HIS CUBS **WILL** DIE.

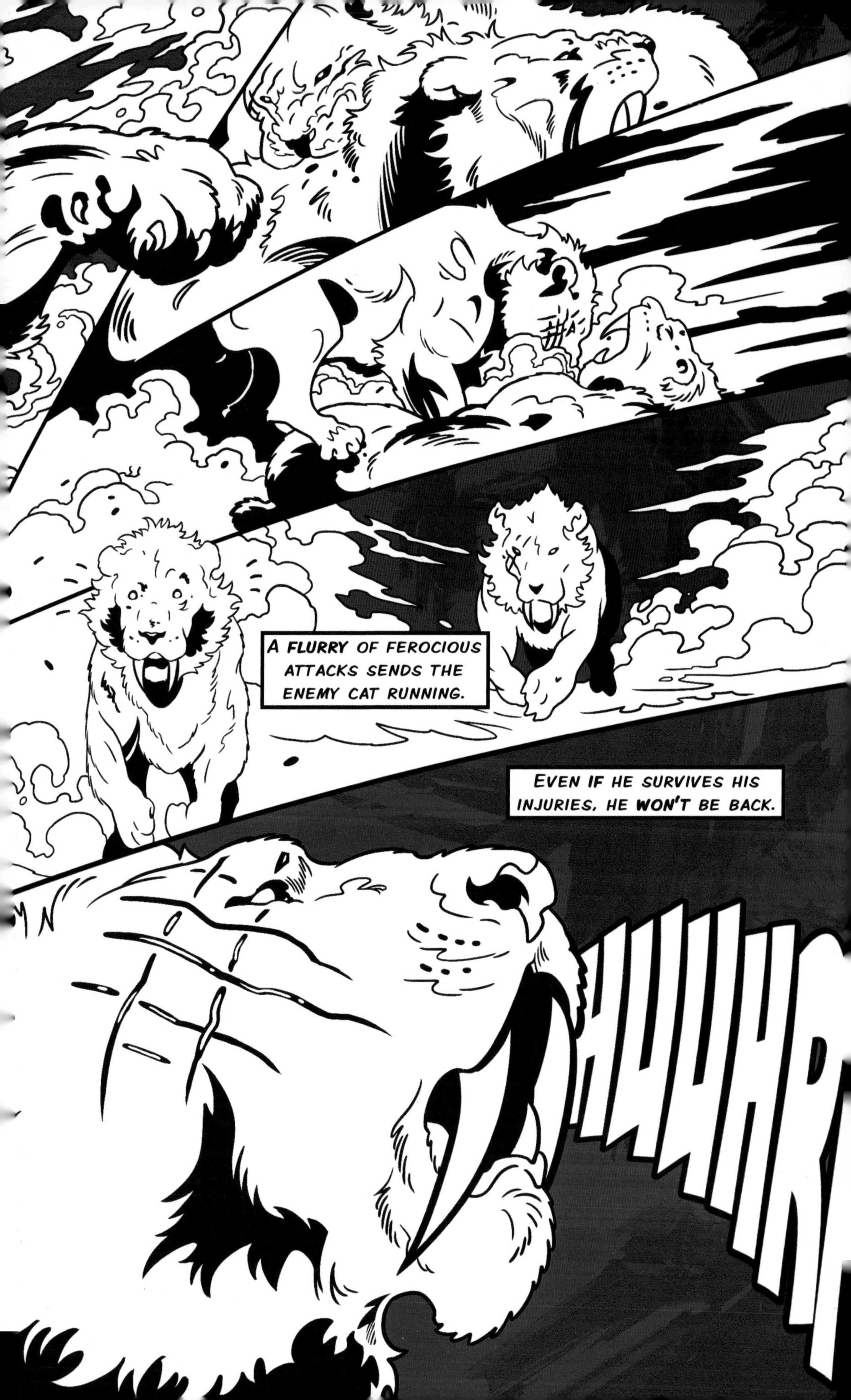
A FLURRY OF FEROCIOUS ATTACKS SENDS THE ENEMY CAT RUNNING.
EVEN IF HE SURVIVES HIS INJURIES, HE WON'T BE BACK.
HUUHR

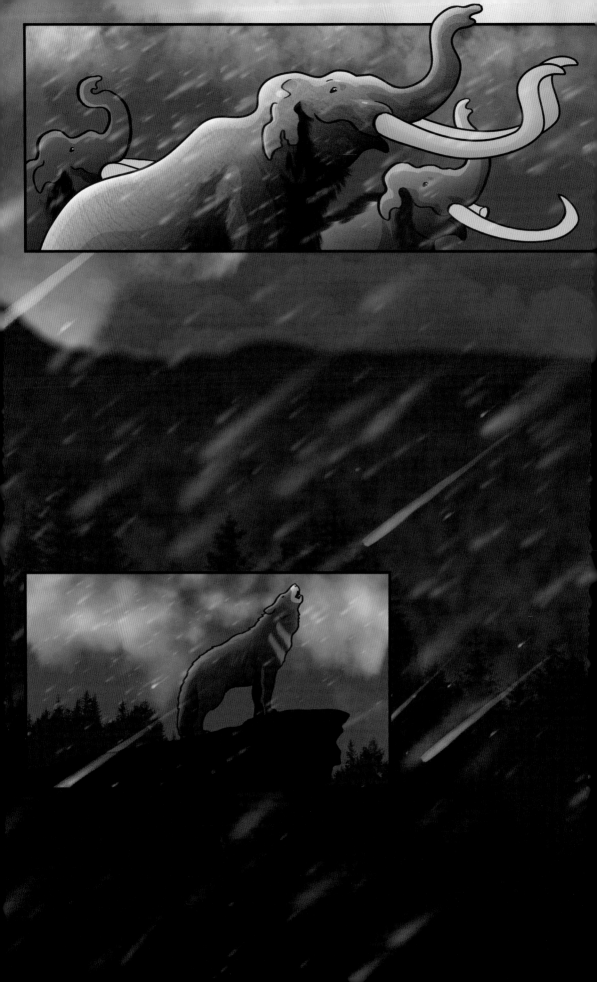

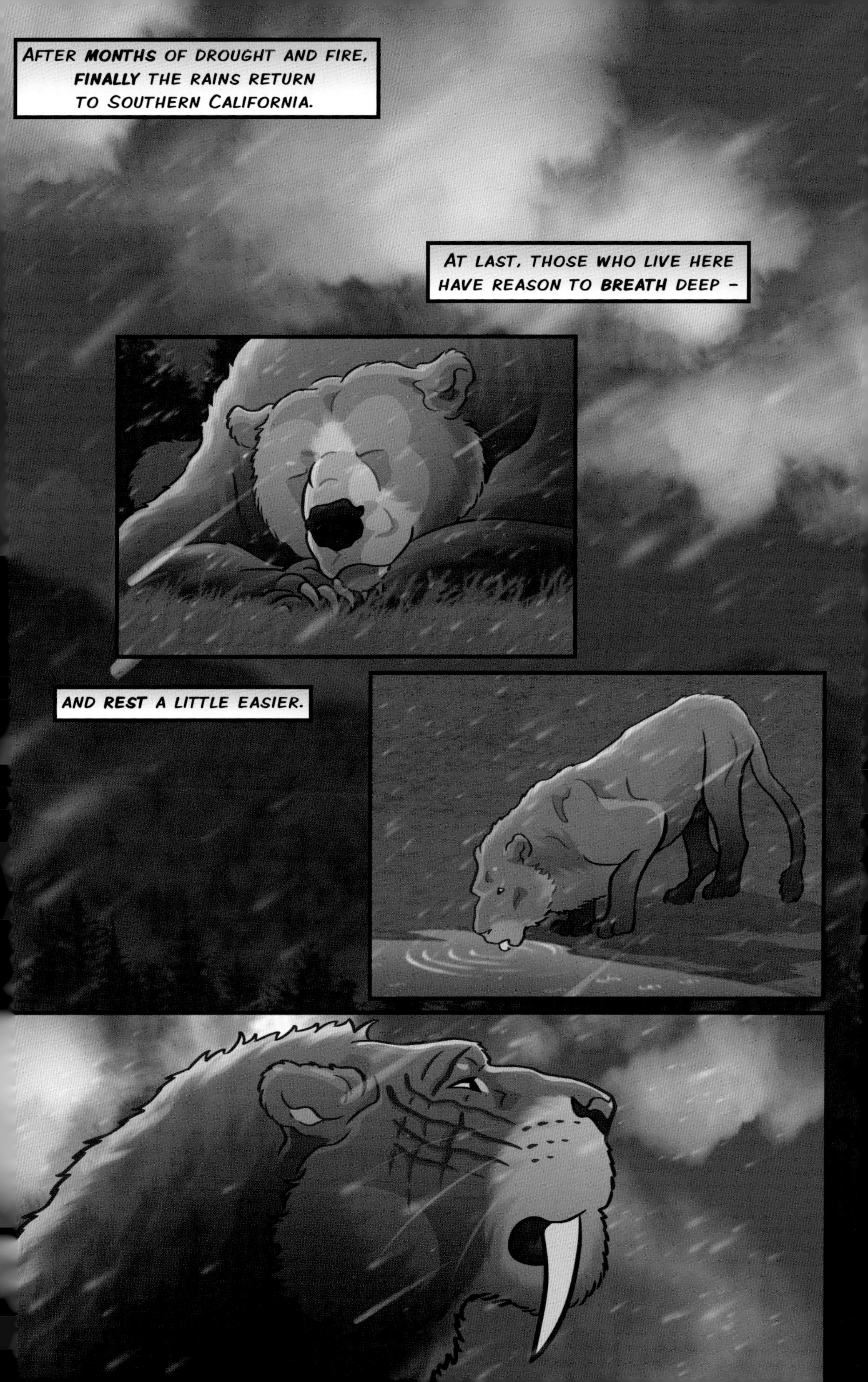
AFTER MONTHS OF DROUGHT AND FIRE, FINALLY THE RAINS RETURN TO SOUTHERN CALIFORNIA.
AT LAST, THOSE WHO LIVE HERE HAVE REASON TO BREATH DEEP -
AND REST A LITTLE EASIER.

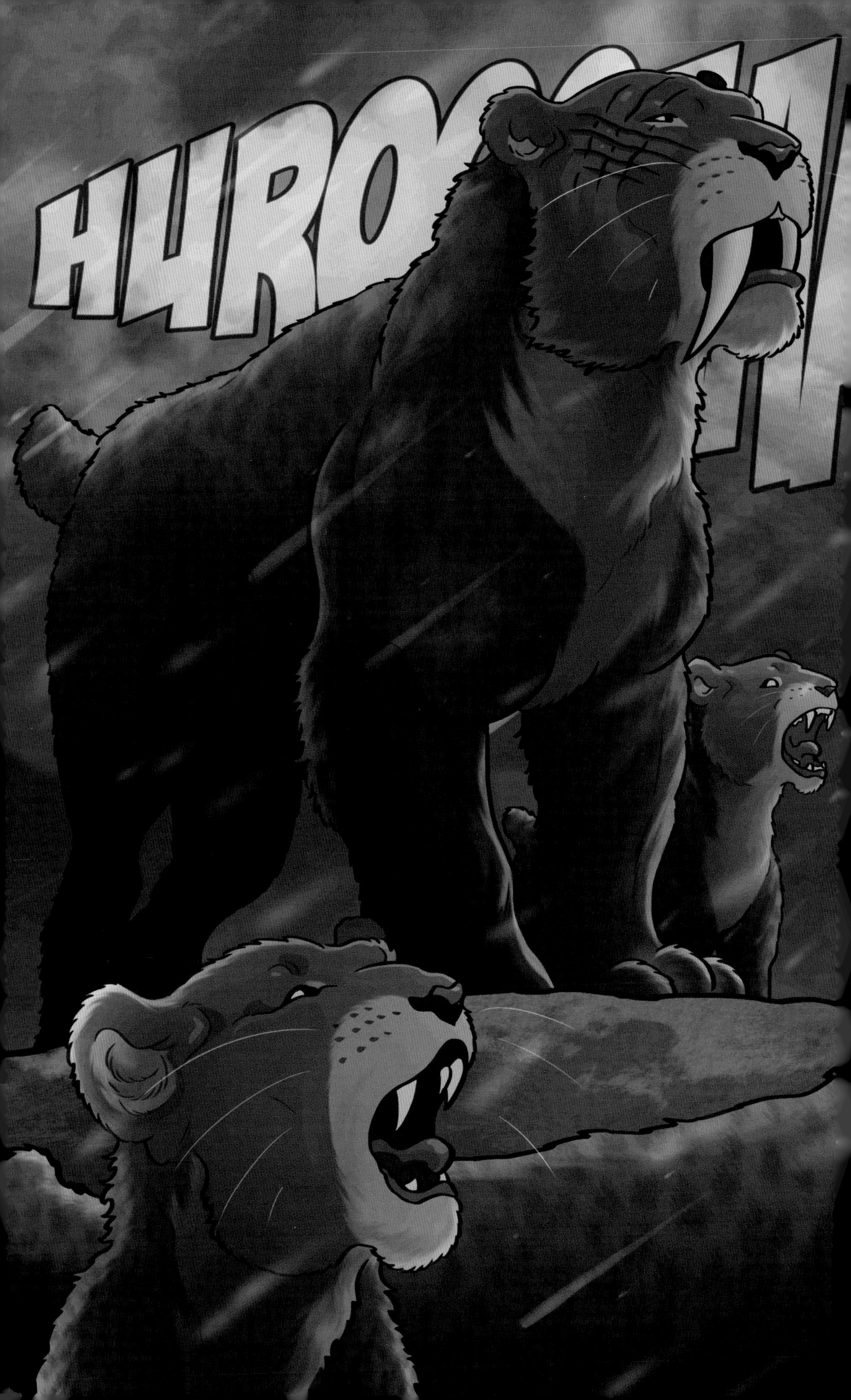
HURO

RRRHHHAAA!
THE RAIN IS SUCH A WELCOME SIGHT, THE SABER CAT FAMILY CAN'T HELP BUT CELEBRATE.
AND THERE'S NO BETTER WAY –
THAN TO ROAR.

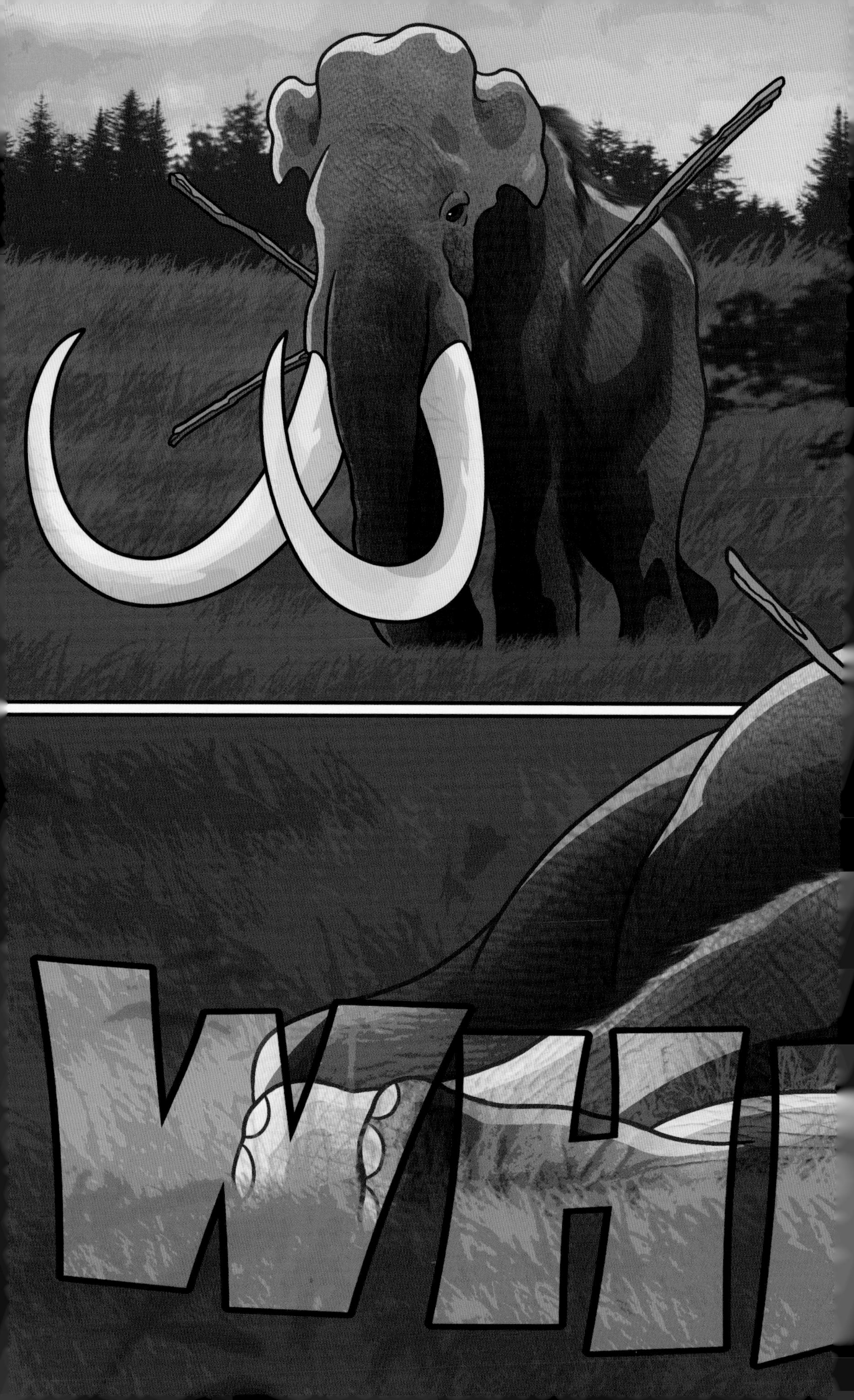
WH

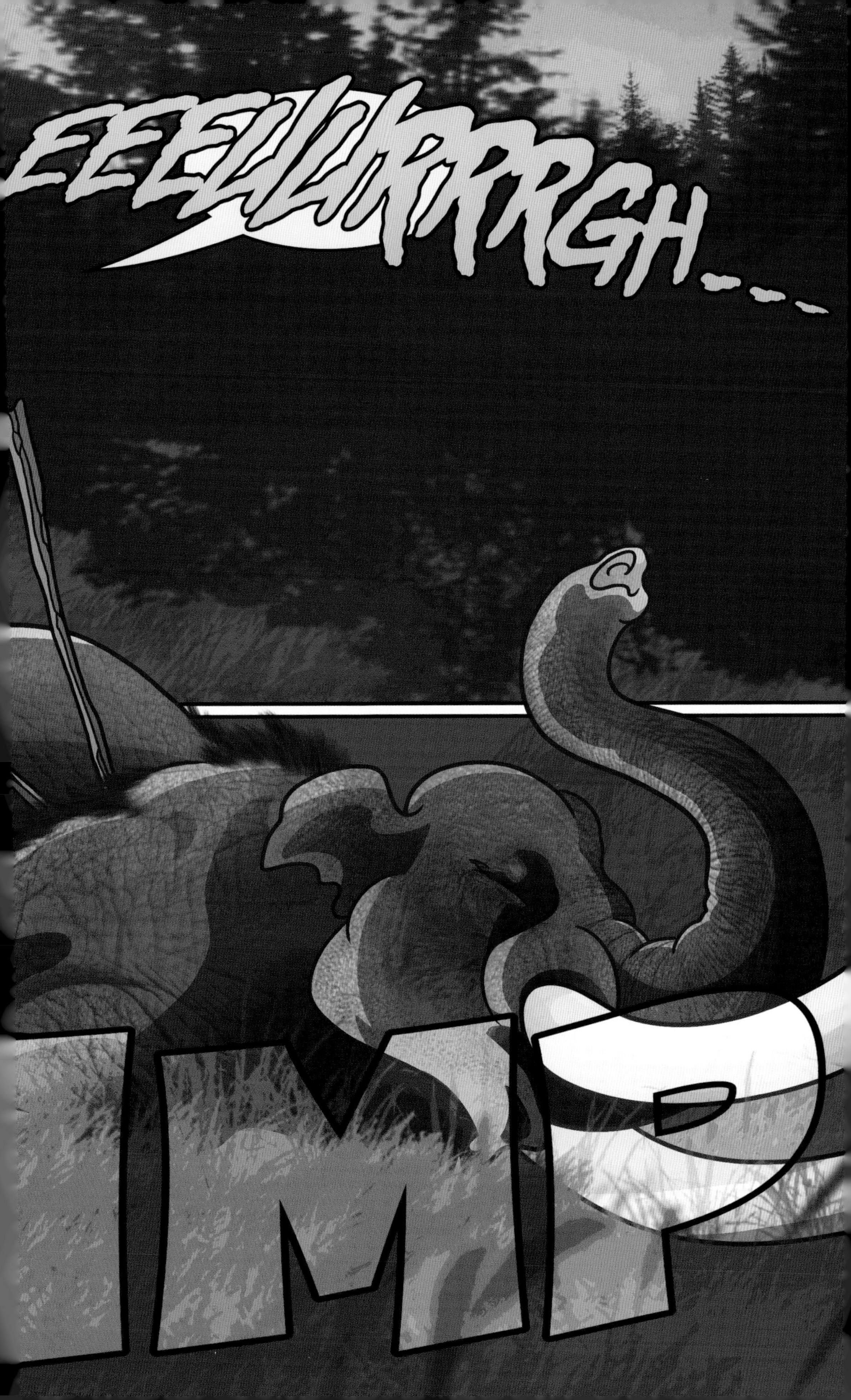
EEEUUUARRGH...
MP

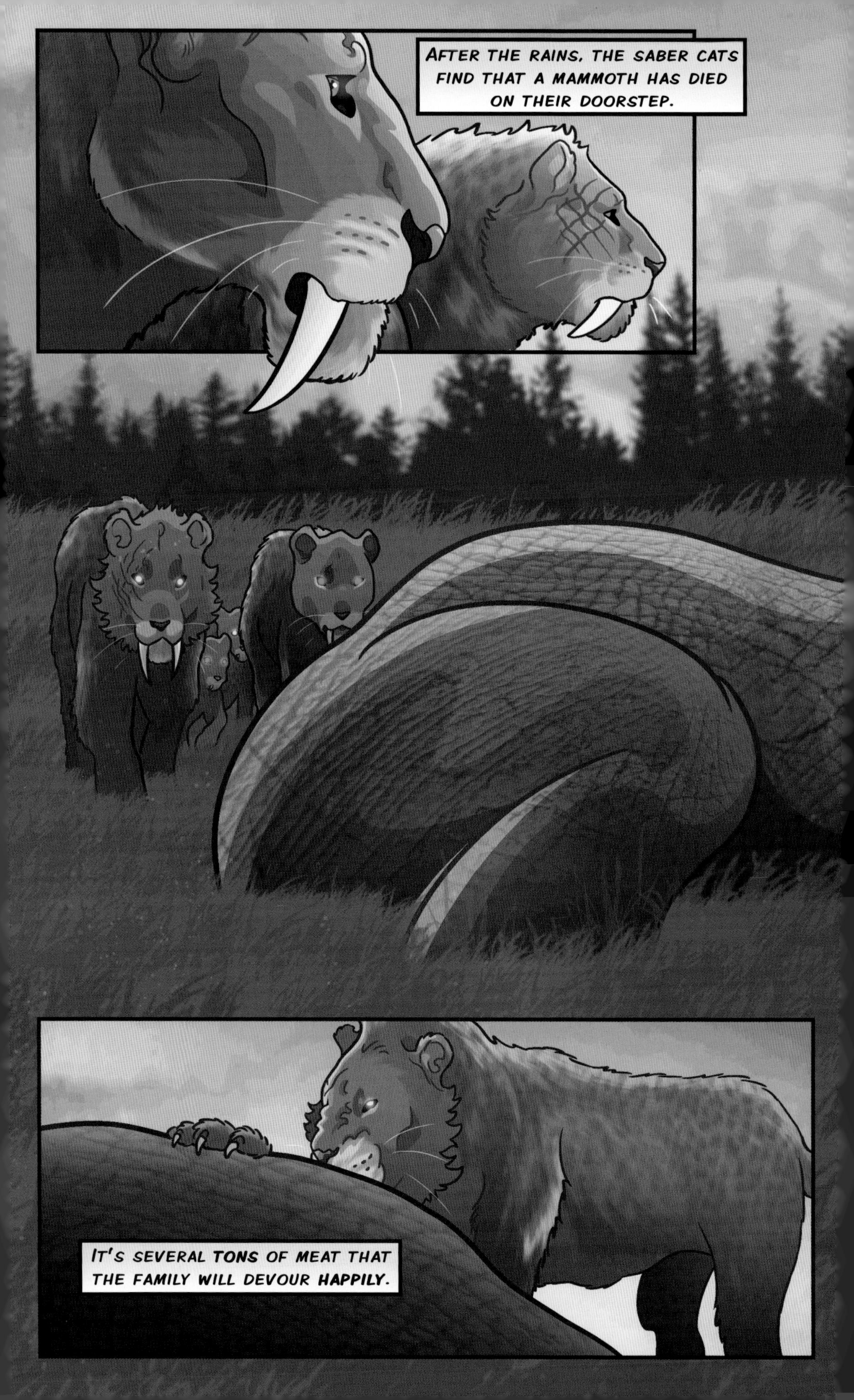
AFTER THE RAINS, THE SABER CATS FIND THAT A MAMMOTH HAS DIED ON THEIR DOORSTEP.
IT'S SEVERAL TONS OF MEAT THAT THE FAMILY WILL DEVOUR HAPPILY.

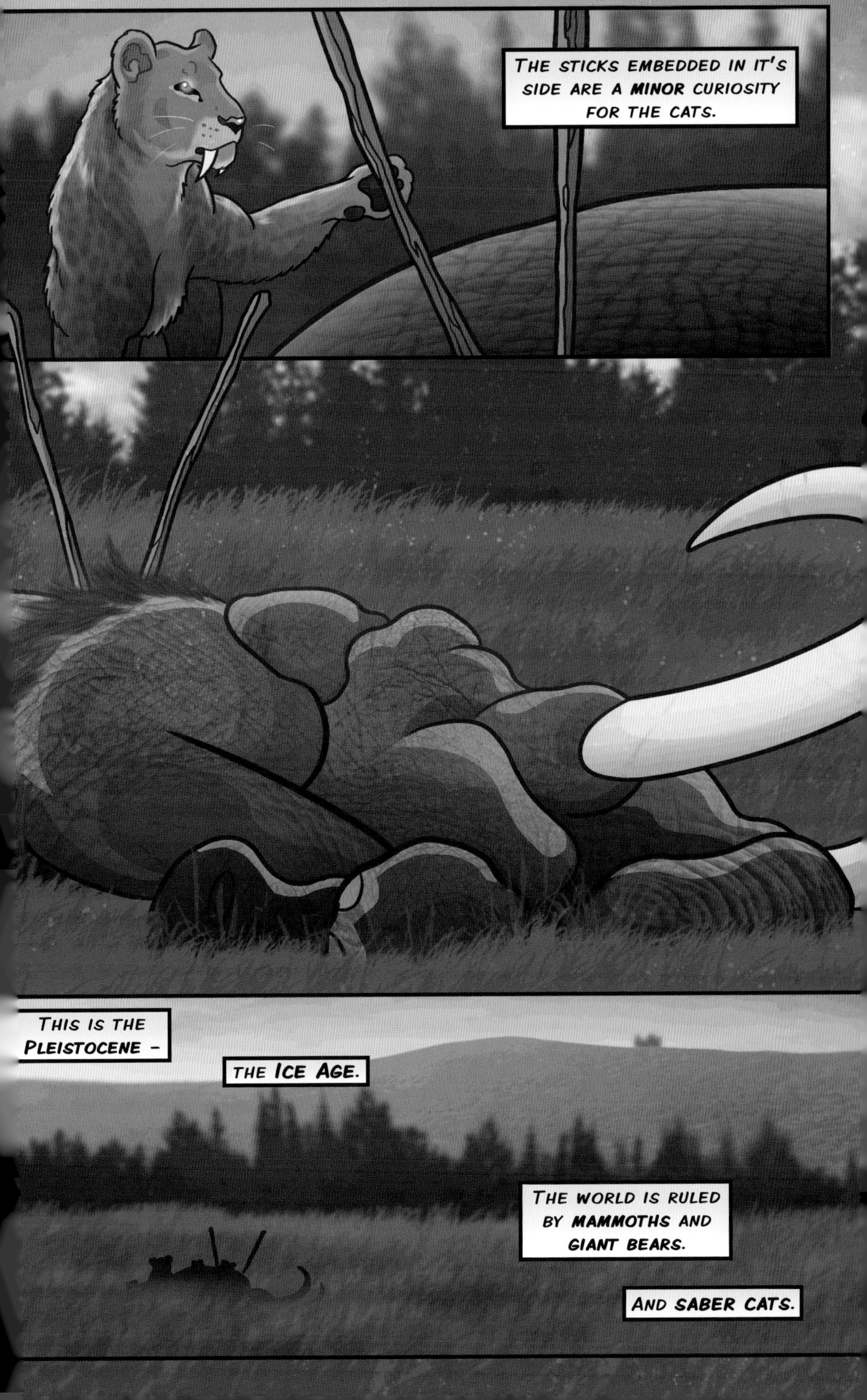
THE STICKS EMBEDDED IN IT'S SIDE ARE A MINOR CURIOSITY FOR THE CATS.
THIS IS THE PLEISTOCENE -
THE ICE AGE.
THE WORLD IS RULED BY MAMMOTHS AND GIANT BEARS.
AND SABER CATS.

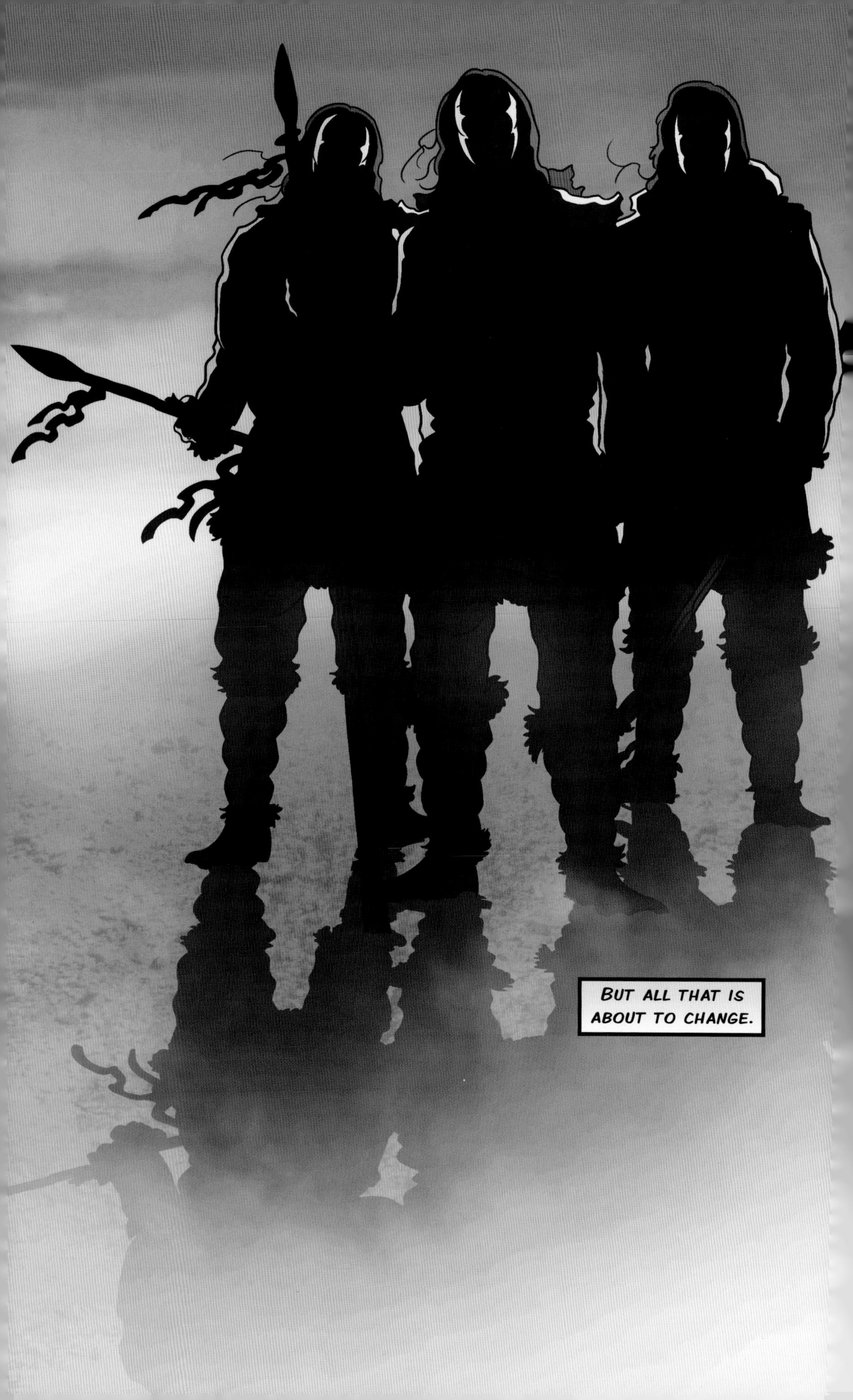
BUT ALL THAT IS ABOUT TO CHANGE.

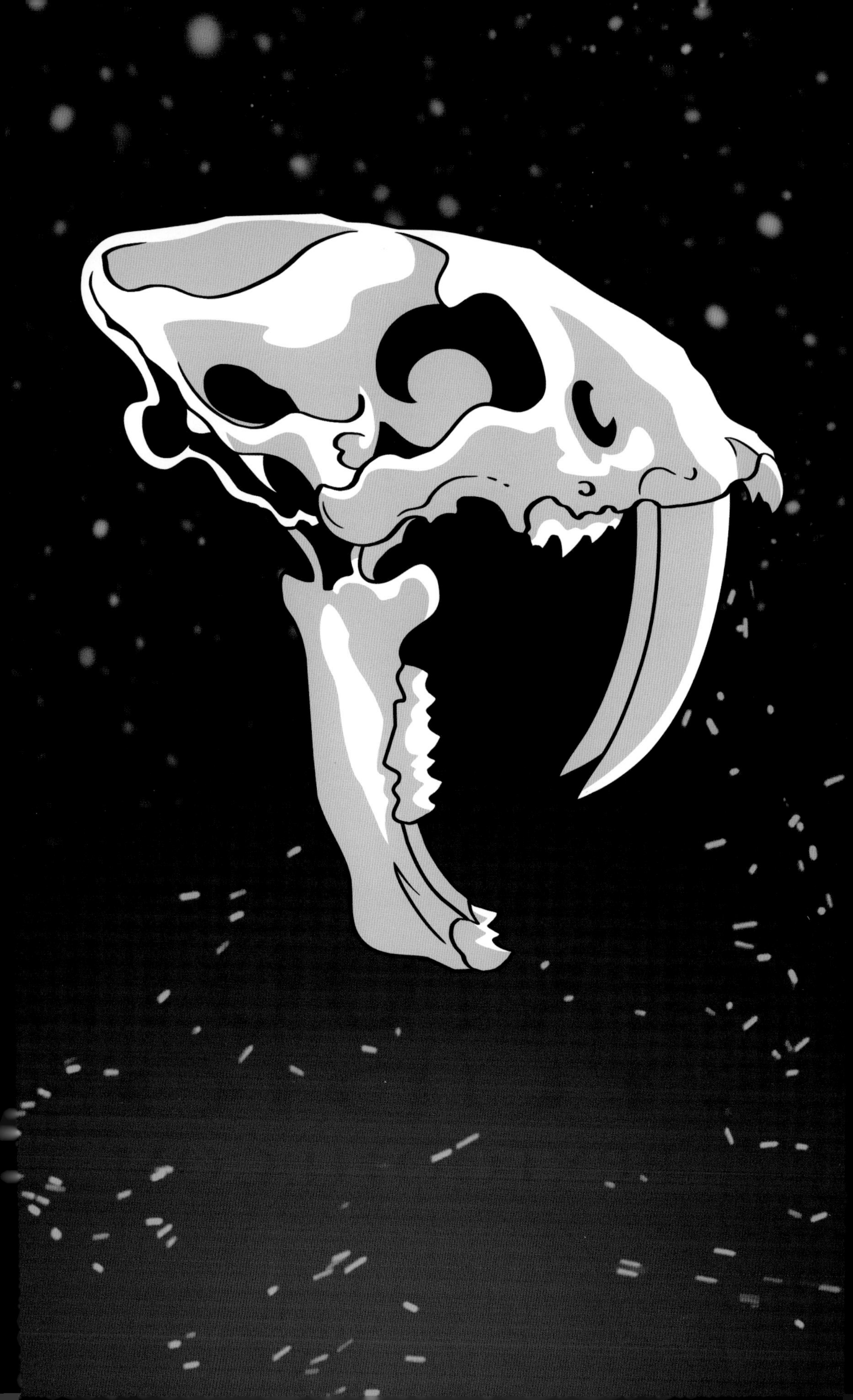

REXTOOTH
STUDIOS
REXTOOTH.COM

MORE BOOKS BY TED RECHLIN

TYRANNOSAURUS REX
TED RECHLIN
FOREWORD BY
@SUEtheTrex

JURASSIC
TED RECHLIN

BEARS
TED RECHLIN

SHARKS
A 400 MILLION YEAR JOURNEY
TED RECHLIN

LOOK FOR THESE TITLES AT YOUR LOCAL STORE, ONLINE, OR AT REXTOOTH.COM

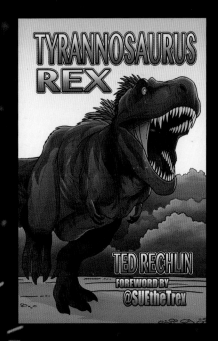

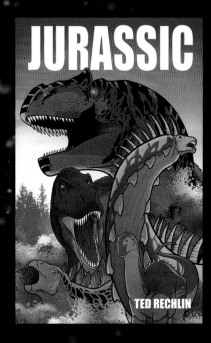

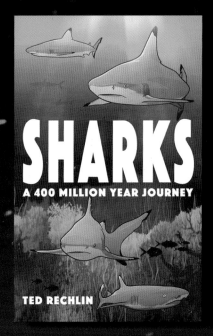

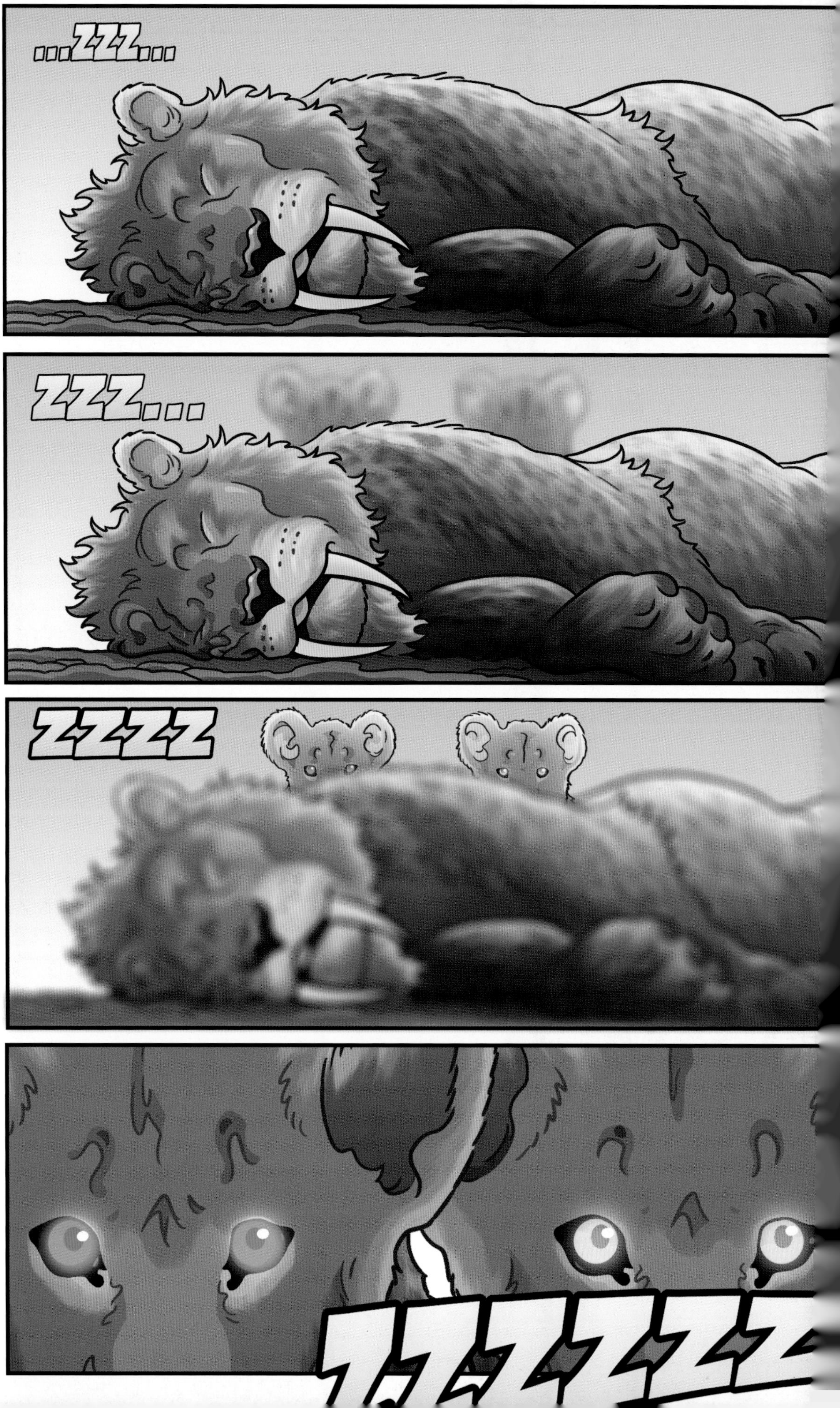
...ZZZ...
ZZZ...
ZZZZ
ZZZZZZ